OXFORD SERIES ON ADVANCED
MANUFACTURING

SERIES EDITORS

J. R. CROOKALL
MILTON C. SHAW
NAM P. SUH

OXFORD SERIES ON ADVANCED MANUFACTURING

3. Milton C. Shaw: *Metal cutting principles*
4. Shiro Kobayashi, T. Altan, and S. Oh: *Metal forming and the finite element method*
5. Norio Taniguchi, Masayuki Ikeda, Toshiyuki Miyazaki, and Iwao Miyamoto: *Energy-beam processing of materials: advanced manufacturing using a wide variety of energy sources*
6. Nam P. Suh: *Principles of design*
7. N. Logothetis and H. P. Wynn: *Quality through design: experimental design, off-line quality control, and Taguchi's contributions*
8. John L. Burbidge: *Production flow analysis for planning group technology*
9. J. Francis Reintjes: *Numerical control: making a new technology*
10. John Benbow and John Bridgwater: *Paste flow and extrusion*
11. Andrew J. Yule and John J. Dunkley: *Atomization of melts for powder production and spray deposition*
12. John L. Burbidge: *Period batch control*

PERIOD BATCH CONTROL

John L. Burbidge

Late of
School of Industrial and Manufacturing Science
Cranfield University

CLARENDON PRESS · OXFORD
1996

Oxford University Press, Walton Street, Oxford OX2 6DP
Oxford New York
Athens Auckland Bangkok Bombay
Calcutta Cape Town Dar es Salaam Delhi
Florence Hong Kong Istanbul Karachi
Kuala Lumpur Madras Madrid Melbourne
Mexico City Nairobi Paris Singapore
Taipei Tokyo Toronto
and associated companies in
Berlin Ibadan

Oxford is a trade mark of Oxford University Press

Published in the United States
by Oxford University Press Inc., New York

© John L. Burbidge, 1996

All rights reserved. No part of this publication may be
reproduced, stored in a retrieval system, or transmitted, in any
form or by any means, without the prior permission in writing of Oxford
University Press. Within the UK, exceptions are allowed in respect of any
fair dealing for the purpose of research or private study, or criticism or
review, as permitted under the Copyright, Designs and Patents Act, 1988, or
in the case of reprographic reproduction in accordance with the terms of
licences issued by the Copyright Licensing Agency. Enquiries concerning
reproduction outside those terms and in other countries should be sent to
the Rights Department, Oxford University Press, at the address above.

This book is sold subject to the condition that it shall not,
by way of trade or otherwise, be lent, re-sold, hired out, or otherwise
circulated without the publisher's prior consent in any form of binding
or cover other than that in which it is published and without a similar
condition including this condition being imposed
on the subsequent purchaser.

A catalogue record for this book is available from the British Library

Library of Congress Cataloging in Publication Data
ISBN 0 19 856400 7

Typeset by Pure Tech Corporation, Pondicherry, India
Printed in Great Britain by
Biddles Ltd, Guildford and Kings Lynn

Coventry University

FOREWORD

Professor John R. Crookall, FEng

I first came to know the name of Professor John L. Burbidge in the early 1960s, when I turned to Production Engineering. It was indeed only a few years before that the first UK university departments of production began to take students. Production Engineering as a university discipline was then new, and as such needed to gain academic respectability, as well as industrial usefulness. Production Engineering as a university subject was contemporary with Management Science, also emerging particularly through Operations Research and other studies, the tendency being to subject the treatment to mathematical rigours. Production Engineering involved the production processes, product engineering, coordination and control through industrial engineering, economics, etc., and these in turn consisted of a collection of mostly numerical techniques aimed at controlling and improving production and its economy.

Meanwhile, Professor Burbidge was striking out for the importance of the overall system of manufacturing. His basic principle was that the simpler the production flow, the more efficient it would be and the easier to manage. One of John's long held objections was that the 'economic batch quantity theory' was misleading, and this book pulls no punches here. However one reaction to his views, I remember, appeared in a prominent paper entitled 'Dragons in pursuit of the EBQ', and John told me years later that he had no doubts that he was the dragon! He went on to write a series of papers entitled 'The case against the economic batch quantity', ... 'against stock control', ... 'against computerised operation scheduling', and so on. Not everyone was pleased. But the true testimony to the breathtaking soundness and industrial applicability of his ideas has come from Japan, through the acknowledged success of the Japanese as a manufacturing nation. Much of that success stems from working practices and methods that had also been espoused by Professor Burbidge. Not surprisingly therefore, John Burbidge was an honoured guest in Japan in his later years. His ideas, comprehensively worked into a major and systematic blueprint for efficient and manageable manufacturing, stand as a key milestone in establishing an eminently practical concept for manufacturing as a whole.

Professor John L. Burbidge OBE (Jack to his friends) was widely honoured for his work. I had both the privilege and pleasure of inviting him at the age of 68 to become Visiting Professor in the College of Manufacturing at Cranfield University, which position he held to the last. He was a teacher greatly respected by the students. Thankfully, he was able to finish this his second book for the Oxford University Press Series on Advanced Manufacturing. Professor Burbidge's earlier book in the Series was *Production Flow Analysis for Planning Group Technology*, published in 1989. He died on 10th January 1995, five days before his 80th birthday. He left four sons, Richard, Peter, David, and Jonathan, and seven grandchildren.

John Burbidge was born in 1915 in Canada of English parents. His father represented R. A. Lister & Co in Canada, and also in the United States. The American Depression forced John to come down prematurely from Cambridge where he had begun his studies. Not disheartened, in 1934 he was apprenticed to the Bristol Aeroplane Company, and as war came he moved to the Ministry of Supply, then joining up as an RAF Engineering Officer. He had memorable tales about the effectiveness of the wartime manufacture of the Spitfire, and looking back this was perhaps the world's first Just-in-Time manufacturing process. He married Dr Elizabeth Newton-Clare on 3rd April 1948. After the war he returned to industry, numbering among his appointments Works Director of British Refrasil Co Ltd and later Managing Director of the Darlington Wire Mills Ltd. This experience laid the foundation for the next stage in 1962, when he became Production Expert with the International Labour Organisation (ILO), serving for five years on assignments in Poland, Cyprus, and Egypt, before becoming Professor in 1967 at the International Training Centre run by the ILO in Turin, from which he retired in 1976.

He had already started to write books and technical papers developing his concepts of production control, including his first article on 'Group Technology', the field with which he was most closely associated, and the subject of his first book in this Oxford University Press Series. 'Group Technology' is essentially the concept of grouping together all the resources necessary to complete 'families' of components and products. This 'product-focused organisation' (Cellular Manufacturing) provides many important advantages, in contrast to the traditional 'process organisation' which groups together similar process operations.

The task of re-designing a factory comprising perhaps hundreds of machines and thousands of parts can be overwhelming, but Professor Burbidge's Production Flow Analysis (PFA) is a systematic method for achieving this re-design and identifying the necessary groupings of machines and components. In 1979, following a programme of research into Group Technology coordinated between the universities of Birmingham, Bradford, Salford, and the London School of Business Studies, and financed by the Science Research Council, Professor Burbidge brought together the results in a book, thereby ensuring coherence and direction towards the engineering industry's real needs. But more than this, his work conducted over some 40 years, during which time fashionable technologies to revolutionise manufacturing have come and gone (along with their financial support), has remained original, forceful in its logic, and always timely. Moreover

Professor Burbidge's ability to encapsulate memorably what those who are interested and discerning recognise to be true has never waned. In all, Professor Burbidge wrote fifteen books and more than 150 papers, many being translated into other languages. At Cranfield he continued to teach his ideas to generations of production engineers at Masters level, ensuring that his fundamental principles of manufacturing will persist. John Burbidge's practice of asking questions during his lectures ensured his students' preparedness to defend their analyses and demonstrate the logic and applicability in practice; but he was always a sympathetic teacher. It was a terrible shock when his wife Betty died in 1985, and he threw himself even more energetically into his work and writings.

Later in his life Jack Burbidge's remarkable worth was widely recognised; he became a Fellow of the Institution of Mechanical Engineers and the British Institute of Management, and was elected an Honorary Fellow of the Production Engineers and Electrical Engineers, and to an Honorary DSc of Strathclyde University, and Novi Sad University in former Yugoslavia. He was active internationally through the organisation for Computer Applications in Production Engineering, and was sought after as a lecturer, giving addresses, even in his last year, in Israel, Ireland, Yugoslavia, and Poland. He was made an OBE in 1991.

Professor Burbidge has stated the indisputable view that 'Japan is today the leading manufacturing nation, because they have mastered the art of low stock manufacturing. The secret of their success in low stock manufacturing is simple material flow, based on continuous line flow, or Group Technology (GT), coupled with Just-In-Time (JIT) production control'. He added that 'to compete with Japan the world must copy them in this'. To demonstrate how to do it is the aim of this book.

PREFACE

Period Batch Control (PBC) is a method of Production Control (PC) developed by the late R. J. Gigli in thirty companies in the UK during the 1930s, and used by him to regulate the manufacture of Spitfire fighter aircraft for the Ministry of Aircraft Production during the Second World War.

Period Batch Control is a single-cycle, flow-control, production control system, which bases ordering on explosion from a series of equal, short-term, production programmes. It can operate successfully at very high rates of stock turnover, and it was one of the earliest just-in-time production control (PC) systems.

When R. J. Gigli was developing PBC in the 1930s as a director of Associated Industrial Consultants Ltd (AIC, the UK successors to Bedaux) Group Technology (GT) had not been developed, and there were no computers in industry. Some of his applications were based on short-term programmes, using one-week periods, but others used four-week periods (with thirteen periods per year). Today, with Group Technology, most applications of PBC are based on one-week periods (fifty per year or more depending on the number of holidays), and much higher rates of stock turnover are achievable than was possible for Gigli.

My first book was published in 1960 with the title *Standard batch control*. It described a method for simplifying production control, by manufacturing assembled products, always in the same, small, standard-size batches. A friend sent a copy to R. J. Gigli who wrote to congratulate me and to invite me to meet him. He said he liked the book, but it would have been better if I had standardized the period rather than the batch quantity. He convinced me then, and I have never had any doubts that he was right since that meeting.

Gigli wrote very little. I have one book, *Material control reference book* (July 1947) published by AIC, which was partly written by him. It mentions, among other topics, his work with the Ministry of Aircraft Production, and gives a rough outline of Period Batch Control. We met several times before he died, and I remember some of the things he told me. I also have a friend, John Christmas, who worked for the same firm of consultants after Gigli died and who was trained to use his methods.

It might be said that the inspiration, the name *period batch control*, and the general philosophy came from R. J. Gigli, but the details of the method described in this book are based on my own experience in using this method in factories.

The main advantages of PBC are: short throughput times, low stocks and its ability to avoid the 'surge effect'. It also reduces obsolescence, eliminates kitting out, and simplifies the measurement of loads and capacity. Period Batch Control was important for Spitfire production because scarce materials, and scarce capacity, were used to make periodic sets of parts for aeroplanes, and not to manufacture parts in large unbalanced batches for stock.

After the Second World War, the use of PBC was gradually reduced. Although it had been used successfully in the 1930s, many engineers after the war saw its insistance on low stocks as another bureaucratic restriction like food rationing, and they were happy to return to what they saw as the safety of high stocks. At this time also, the Americans developed Materials-Requirement Planning (MRP), which is normally operated as a multicycle flow-control system, and which was originally based on the use of so-called economic (*sic*) batch quantities.

Materials-Requirement Planning (MRP) did a great service to production by introducing computers for routine office work in PC, but when the Japanese introduced just-in-time PC with simple materials-flow systems (MFSs), coupled with the reliability of Total Quality Control, companies with MRP and functional organization were unable (and are still unable) to match the high rates of stock turnover achieved with GT and SIT, or to gain the major economic advantages which this gave them. We now need to return to PBC before it becomes a lost art.

Part 1 of this book provides an introduction to the subject. Chapter 1 provides an outline of PC, and Chapter 2 describes Group Technology (GT). It might be said that in most factories, GT only achieves its full potential if it is coupled with PBC. Conversely, PBC with short periods of one week or less only works efficiently if it is based on GT.

Chapter 3 describes the feedback controls used to constrain events to follow the plans made in production control, and Chapter 4 is about stocks and the factors which determine the stock levels in factories.

Part 2 describes the planning of a PBC system for a company, and how such a system is operated in practice. Seven chapters cover: the data bank for PBC, programming, ordering, dispatching, and operation scheduling. One of these chapters, Chapter 9, is about the time constraints, such as the load on bottleneck machines, throughput times and set-up times, which may have to be reduced before PBC can be used, and studies the relatively simple and inexpensive ways by which this can be done.

Part 3 is about the application of PBC in different types of industry. Different chapters deal with PBC in implosive industries (foundries, potteries, and glass works, for example), in explosive (assembly) industries, in process and square industries, and in jobbing factories.

Finally, Part 4 deals with the problem of introducing PBC in a factory, and with the methods of controlling its introduction.

This book ends with an Index and a Glossary of terms. The Glossary gives the meanings that I ascribe to the principal terms used in this book.

This is necessary because of the confusion caused by the recent tendency of engineers to 'reinvent the wheel' and then change its name. Important examples are the changes which introduced *manufacturing* to replace *production*, *cellular production* to replace *group technology*, and *logistics* to replace *production control*. The danger is that by changing the name we will lose the line of development. In general I have preferred the traditional terminology to the new buzz-words.

Cranfield
February 1994

J. L. B.

CONTENTS

PART 1 INTRODUCTION

1 An outline of production control   3
   1.1 Introduction   3
   1.2 Programming   3
   1.3 Flow-control or stock-base ordering systems   6
   1.4 Ordering systems single-cycle and multicycle   11
   1.5 Just-in-time or just-in-case ordering systems   12
   1.6 The economic-batch quantity   14
   1.7 Dispatching   18
   1.8 Summary   21

2 The materials-flow system and Group Technology   22
   2.1 Introduction   22
   2.2 Elements of GT   24
   2.3 GT stages   25
   2.4 The advantages of GT   29
   2.5 Planning the division into groups and families   29
   2.6 FFA and GA   31
   2.7 Assembly groups   34
   2.8 Materials transfer between GT-component-processing and assembly groups   36
   2.9 Total Group Technology   41
   2.10 Summary   41

3 The production-control feedback controls   42
   3.1 Introduction   42
   3.2 Systems theory   43
   3.3 Connectance   44
   3.4 The design of the system   46
   3.5 Progressing   47
   3.6 Loading   53
   3.7 Summary   55

4 Stocks   57
   4.1 Introduction   57
   4.2 Stocks   57
   4.3 Cycle stocks and related variables   60
   4.4 Reducing cycle stocks   61
   4.5 Buffer stocks   65
   4.6 Excess stock   71
   4.7 Dynamic changes in stocks   73
   4.8 Inventory control   74
   4.9 Summary   77

PART 2 PROGRAMMING, ORDERING AND DISPATCHING WITH PBC

5 The database for PBC 81
5.1 Introduction 81
5.2 The plant list 81
5.3 The list of employees 84
5.4 Process routes 86
5.5 Parts lists 88
5.6 Operation sheets 89
5.7 Lists of suppliers 90
5.8 Lists of customers 90
5.9 Maintaining the accuracy of the database 91
5.10 Summary 93

6 The annual programme for PBC 94
6.1 Introduction 94
6.2 Planning the annual sales programme 97
6.3 The method of statistical forecasting 98
6.4 Trends 98
6.5 Seasonal variations 102
6.6 Product life cycle 104
6.7 Cyclical variations 104
6.8 Programme reconciliation 106
6.9 The annual production programme 106
6.10 The finished-product stock programme 106
6.11 Summary 107

7 Short-term or flexible programming for PBC 108
7.1 Introduction 108
7.2 Selecting the programming period 109
7.3 Selecting the programming schedule 109
7.4 Standard assembled products and their programming schedule 111
7.5 Products sold as piece-parts 119
7.6 Jobbing products 121
7.7 Short-term programming for process industries 122
7.8 The programme meeting 123
7.9 Summary 124

8 Ordering and purchasing with PBC 125
8.1 Introduction 125
8.2 Purchasing 125
8.3 Ordering materials for process and implosive industries 127
8.4 Ordering standards for assembled products 129
8.5 The ordering schedule 129
8.6 Summary 133

9 The time constraints and PBC 135
9.1 Introduction 135
9.2 Capacity and load 136
9.3 Throughput time 139

CONTENTS xiii

	9.4 Set-up time	143
	9.5 The sequence-constraint problem	149
	9.6 Purchase lead times	151
	9.7 Summary	151
10	**Dispatching with PBC**	**152**
	10.1 Introduction	152
	10.2 Setting	153
	10.3 Tooling storage and maintenance	156
	10.4 Inspection	159
	10.5 Materials handling	160
	10.6 Plant maintenance	160
	10.7 Swarf removal	160
	10.8 Good housekeeping	162
	10.9 Records	162
	10.10 Summary	162
11	**Operation scheduling with GT and PBC**	**163**
	11.1 Introduction	163
	11.2 The scheduling problem	163
	11.3 Factors in planning the launch sequence	164
	11.4 Factors affecting the reliability of LSS	170
	11.5 Planning the component launch sequence for a group	173
	11.6 Flexibility with LSS	174
	11.7 Summary	175

PART 3 PBC IN PRACTICE

12	**PBC for implosive industries**	**179**
	12.1 Introduction	179
	12.2 The product	179
	12.3 The method of manufacture	180
	12.4 The annual programme	180
	12.5 The short-term programmes	180
	12.6 Ordering materials	183
	12.7 Ordering in other implosive industries	186
	12.8 Summary	188
13	**PBC in explosive industries**	**189**
	13.1 Introduction	189
	13.2 The range of products produced	189
	13.3 The materials-flow system	190
	13.4 Planning PBC for the factory	191
	13.5 PBC in a machine-tool factory	197
	13.6 The advantages of PBC	198
	13.7 The problems with PBC	198
	13.8 Summary	199
14	**PBC in process and square industries**	**200**
	14.1 Introduction	200
	14.2 Fixing the capacity and Output levels	200

14.3	Process industries, seasonal products	202
14.4	A process industry with an even demand	203
14.5	PBC in industries making products with a short shelf-life	204
14.6	PBC in square industries	208
14.7	Summary	209

15 PBC in jobbing factories 210

15.1	Introduction	210
15.2	Make-ready	210
15.3	Planning PBC for jobbing	211
15.4	Programming	213
15.5	Ordering	214
15.6	Examples from practice	214
15.7	PBC with large one-of-a-kind projects	217
15.8	Summary	217

PART 4   INTRODUCING PBC

16 Introducing PBC 221

16.1	Introduction	221
16.2	Objectives	222
16.3	Preliminary changes	223
16.4	Planning the series of change projects in explosive industries	227
16.5	Planning the change projects in other industries	230
16.6	Summary	231

17 Controlling PBC introduction and operation 232

17.1	Introduction	232
17.2	Controlling the introduction of PBC	232
17.3	The schedule for introduction	233
17.4	The investment in the change to PBC	235
17.5	Savings and benefits	239
17.6	Control during operation	241
17.7	Summary	241

18 Conclusion 243

18.1	Introduction	243
18.2	The materials-flow system	243
18.3	Simplifying production control	245
18.4	Period Batch Control	247
18.5	Economic savings with GT plus PBC	248
18.6	The social benefits of GT and PBC	249
18.7	Conclusion	252

GLOSSARY 253

INDEX 260

PART 1

INTRODUCTION

1

AN OUTLINE OF PRODUCTION CONTROL

1.1 Introduction

Production Control (PC) is a function of management, and Period Batch Control (PBC), the subject of this book, is one particular method of PC. In order to see how PBC compares with other established methods, this chapter summarizes the whole field of production control.

The regulation by PC of the flow of materials through the Material-Flow System (MFS) of an enterprise takes place at the three main levels of programming, ordering, and dispatching. The output at these levels is planned or, more exactly, is *scheduled* progressively, as shown in Fig. 1.1. Scheduling, in general, plans the starting and/or finishing times for work tasks. *Programming* schedules the output of finished products from the factory. *Ordering* schedules the output of component parts from the factory workshops and the input of bought parts and materials from suppliers. Finally, *dispatching* schedules the completion of the operations done to complete orders in the factory workshops.

Three main feedback controls are used in PC to constrain events to follow plans. These controls *monitor*, or record, the actual achievement and compare it with the planned achievement. This is followed by the *feedback* of information to management about significant variances between them. The three controls are:

(1) *progressing*, which controls the product, component, and operation output at the three PC levels;
(2) *loading*, which controls the capacity and the load in the factory;
(3) *inventory control*, which controls the level of stocks.

This chapter gives a broad analysis of the three PC levels, and the three feedback controls are studied in Chapters 3 and 4. It treats the batch production of standard mechanical assembled products as the general case, and other types of industry are treated as variants which may require variants of the PBC production control method.

1.2 Programming

A programme (or a master production schedule in the USA) is a schedule for the completion of products. The *term* of a programme or its planning horizon

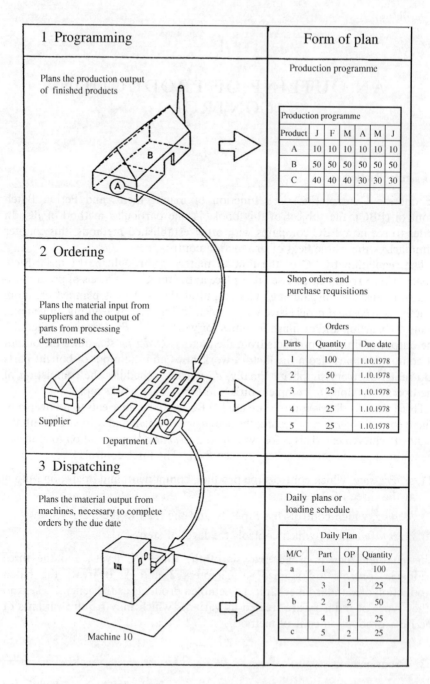

FIG. 1.1 Levels of production control. (Reprinted with the permission of William Heinemann Ltd.)

in the USA is the time it covers into the future. Programmes are generally needed covering three main terms:

(1) *long term*, issued (say) yearly, covering a term of five years or more, and used for long-term corporate planning;
(2) *annual*, issued yearly, covering a term of one year, and used for financial planning and control, and for other medium-term plans, including the planning of purchase contracts;
(3) *short term*, newly issued every period, covering one or a small number of periods into the future, and used to regulate assembly and purchase deliveries on *call-off* and in Period Batch Control PBC to regulate the component output from workshops this short-term programming method is also called *flexible programming*.

In the case of the annual and short-term programmes, a set of three complementary programmes is generally needed, covering:

(1) a sales programme (sales deliveries to customers);
(2) a production programme (the finished product completions planned for the factory);
(3) a stock programme (the finished product stock at the end of each period).

A set of such programmes is illustrated in Fig. 1.2. The annual sales programme is a forecast which is normally based on statistical forecasting and on market research. It shows the planned deliveries to customers in a series of future periods (for example, months, two-week periods, or weeks) covering one year. The annual production programme attempts to meet the needs of the annual sales programme, and to *smooth* random and seasonal variations in demand at the same time, and thus to save capacity by making additional products for stock when the sales demand is below the mean demand rate. These smoothing stocks are used to supplement new production when the sales are above the mean demand rate. The *finished-product-stock* programme is found by direct calculation from the sales and production programmes.

The short-term sales programmes next, are based on a series of short periods, typically of one week. They cover a fixed term of a small number of periods

Programme	J	F	M	A	M	J	J	A	S	O	N	D	Total
Sales	10	10	10	17	20	20	20	20	15	15	15	15	187
Production	16	16	16	16	16	16	16	15	15	15	15	15	187
Stock O	6	6	18	17	13	9	5	0	0	0	0	0	

FIG. 1.2 A set of programmes.

per cycle, specified in a standard programming schedule. Two main methods are used for planning short-term sales programmes:

(1) periodic accumulation of orders;
(2) periodic review.

The periodic-accumulation method accumulates the sales orders received in each period, and this list is issued as the sales programme, showing the required product deliveries to customers in some period at a specific small number of periods into the future. This method is generally associated with standard products and with a fixed delivery promise time.

The periodic-review method is used when a range of products is produced with many design variants, with different lead times and delivery promise times. The programmes in this case are generally fixed at a programme meeting each period which is attended by delegates from sales, production, and the other disciplines affected. The delegates will try to plan a set of programmes which ensures that all the products are delivered by the promised date and at the same time 'smooths' production to give an even rate of production.

The main purpose of a production programme is to regulate the completion of finished products. Figure 1.3 shows a classification of manufacturing industries based on the ratio of the number of material types, m, to the number of finished-product types, p. In the process industries, 'm' and 'p' are both small; in the implosive industries, 'm' is very small and 'p' is large; in the square industries, 'm' and 'p' are equal and large; and, in the explosive industries, 'm' is large and 'p' is relatively small. In the explosive industries, the production programmes will be assembly programmes for finished assembled products. In the process, implosive, and square industries—which make bulk materials or parts rather than assemblies—the production programme will take the form of lists of these items showing the quantities needed. The programme in this case also constitutes an *order* for parts manufacture. The programming and ordering levels are combined in these cases as far as manufacturing is concerned.

1.3 Flow-control or stock-base ordering systems

Ordering systems, at the second level of PC can be classified as either *flow control* or *stock-base* systems.

Flow-control systems all start with a production programme, followed by *requirement scheduling*, or calculation of the quantities of parts needed to complete the given programme (see Fig. 1.4). They then continue in a variety of different ways to regulate the issue of orders to meet the requirement schedule. Examples of flow-control systems include component scheduling, Materials-Requirement Planning (MRP), Period Batch Control (PBC) and Optimum Production Technology (OPT). These systems are also illustrated in Fig. 1.4.

	Process	Implosive	Square	Explosive
m = Material varieties p = Product varieties	$m = p$	$m \triangleleft p$	$m \square p$	$m \triangleright p$
Material input	Bulk material	Bulk. or gen. material	Components	Gen. or Sp. mtl.
Product output	Bulk material	Gen. mtl. or components	Components	Assemblies
Material flow type	Line flow	Batch or line flow	Batch flow	Batch flow
Examples	B Cement E Ore treatment F Milk F Sugar C Gases, e.g. O_2, N B Bricks B Timber F Breweries E Tanneries E Paper	E Foundries D Potteries D Glass C Decorated laminates T Spinning, fibres E Brake linings D Printing T Knitting F Bakeries E Rolling mills E Wire drawing	E Jobbing machining (some) D Dry cleaning E Machine overhaul T Dyeing of textiles T Finishing of textiles E Heat treatment E X-ray E Painting E Electroplating E Metal spraying E Polishing	E Automobile E Electronics D Consumer durables E Machine tools E IC engines E Electrical T Weaving D Clothing and shoes C Chemical dye stuffs D Furniture E Welding

Key: F = Food. C = Chemicals. T = Textiles. E = Engineering. B = Building. D = Domestic. IC = Internal combustion. Gen. mtl. = General materials. Sp. mtl. = Special materials.

FIG. 1.3 Types of Industry. SCC classification: F, food; C, chemicals, T, textiles; E, engineering; B, building; and D, domestic.

A. Common to all flow-control systems, calculate the requirement schedule from the programme for product X.

Month	J	F	M	A	M	J	J	A	S	O	N	D
Programme	10	12	14	20	20	20	18	14	14	12	16	16
Part 1	10	12	14	20	20	20	18	14	14	12	16	16
Part 2		20	24	28	40	40	40	36	28	28	24	32	32
Part 3		10	12	14	20	20	20	18	14	14	12	16	16
etc.													

B. The four flow-control systems, with different scheduling methods, from the same requirement schedule:

(1) Component scheduling – Advances long-term programme to provide lead time.

Month	D	J	F	M	A	M	J	J	A	S	O	N	D
Part 1	10	12	14	20	20	20	18	14	14	12	16	16	14

(2) MRP   Explodes from long-term programme. Different for each part W

Part 1

OQ = 60   Lead time, L = 2 months   ⊖ = Order date

(3) PBC   Explodes from short-term programme (P) to find (M)

Week number	1	2	3	4	5	6
Cycle 1	●	O	M	P	S	
Cycle 2		●	O	M	P	S
Cycle 3			●	O	M	P	S

Key:
● = Programme meeting
O = Get materials
M = Make parts
P = Assemble
S = Deliver to customer

(4) OPT Similar to PBC, but O = Operation schedule on computer. Schedule for other machines, based on schedule for bottlenecks.

FIG. 1.4 Flow-control ordering systems.

A problem with this classification of ordering systems, is that the term Materials-Requirement Planning (MRP) is also a synonym in the USA for requirement scheduling. Since the term MRP covers both requirement scheduling and also the rest of the MRP ordering system, the Americans tend to see PBC and OPT as variants of MRP and not as different types of ordering system.

Stock-base ordering systems regulate the issue of orders in response to changes in component stock levels. There are three main types of stock-base systems:

1. *Stock-Control (SC) systems* use a fixed order quantity, and a fixed reorder level in a declining stock (at which new orders must be issued).

2. *Stock-Replacement (SR) systems* These have a fixed stock level for every item. Any issues from stock are covered by replacement orders, issued either immediately after the issue of a container load (Kanban), or at the beginning of the next period (PBC).

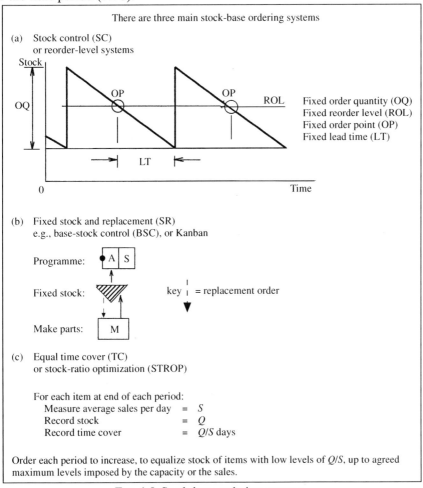

FIG. 1.5 Stock-base ordering systems.

3. *In Time-Cover (TC) systems*, the sales rate is recalculated at regular intervals for all products. The stock divided by the sales rate gives the *time cover* of the stocks. Orders are issued to raise and equalize the time cover for items with the least cover. The method of doing this is called *Stock-Ratio Optimization* (STROP).

Figure 1.5 illustrates these different stock-base systems. The stock-replacement and time-cover systems have a future, but the stock-control system is probably obsolete for regulating the provision of parts for manufacturing—this is for a number of reasons, not least because it is a multicycle system and is subject to the surge effect (see the next section). Another special problem with stock-control systems is that if they are used to transmit demand between a series of independent inventories, such as from retailer to wholesaler to factory stock, they generate a progressive magnification in the demand variation (see Fig. 1.6). This effect has been widely observed in industry and was

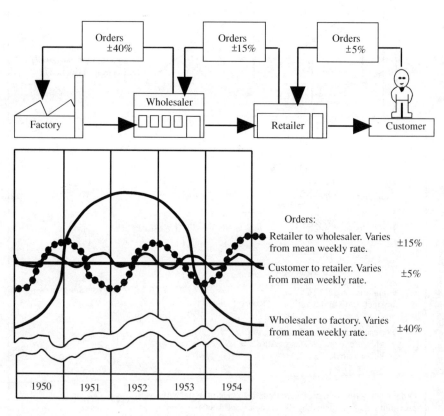

FIG. 1.6 The industrial-dynamics effect.

AN OUTLINE OF PRODUCTION CONTROL

demonstrated by Jay W. Forrester in his research at the Massachusetts Institute of Technology (MIT) entitled *Industrial dynamics*.

1.4 Ordering systems single-cycle and multicycle

A second method of classifying PC ordering systems divides them into single or multicycle systems, as illustrated in Fig. 1.7. With single-cycle systems, all orders are issued on a series of common order dates (OD) at regular period intervals, for completion by a series of common due dates at the same intervals. With multicycle systems, different parts are ordered to different order frequencies (orders per year), these were traditionally chosen on the basis of the so-called economic-batch quantity (EBQ) theorem.

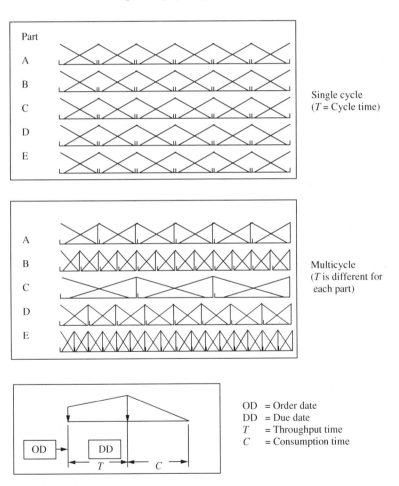

FIG. 1.7 Single and multicycle ordering.

Both stock control and MRP are multicycle systems. Kanban, Period Batch Control (PBC) and OPT on the other hand, are single-cycle systems at the programming level. Period Batch Control (PBC) is also single cycle at the ordering level. Optimum production technology (OPT) is more concerned with machines than parts, giving preference in scheduling to the bottleneck machines. The Kanban system orders parts to replace those taken from stock immediately after containers of parts are issued to the next process. It is therefore also indirectly a single-cycle ordering system if the final process (assembly) is regulated by short-term programming.

Multicycle systems have major deficiencies which make them unsuitable for use as ordering systems, and which make the traditional stock-control and MRP systems obsolete.

The first deficiency is that because multicycle systems generally cover a wide range of order frequencies they have to 'explode' to find requirement schedules for ordering parts, from long-term programmes. These will include some parts with low order frequencies, covering several months into the future. As it is since not given to human beings to foretell the future, these long-term programmes have to be revised several times a year, leaving part-worked batches of parts which are no longer needed and require special rush orders for any replacement parts.

The second deficiency is that multicycle systems are subject to the *surge effect*, which makes them unstable. As illustrated in Fig. 1.8, the peaks and troughs of the different component-stock cycles and load cycles drift into and out of phase with time, producing unpredictable variations in the total stock and in the total load.

The third deficiency is that these multicycle systems are very complicated in comparison with single-cycle systems, because each part has to be scheduled individually. A special deficiency in the case of assembly (explosive) industries is that the collection of sets of parts for assembly (kitting out) is more unreliable, time-consuming, and expensive for multicycle ordering systems than is the case for single-cycle systems.

1.5 Just-in-time or just-in-case ordering systems

A third classification of ordering systems divides them into *just-in-time* (JIT) and *just-in-case systems*. The just-in-case category covers most of the traditional ordering methods, which tend to manufacture in large batches and to carry heavy stocks (just-in-case they might be needed).

The "just-in-time" (JIT) category covers a class of PC ordering methods of growing importance. In its latest form, the JIT method was developed by the Toyota car company in Japan, but a very similar philosophy was adopted by R. J. Gigli for PBC in the 1930s. It said:

1. With minor reservations for smoothing, products should not be produced until they can be delivered to customers.

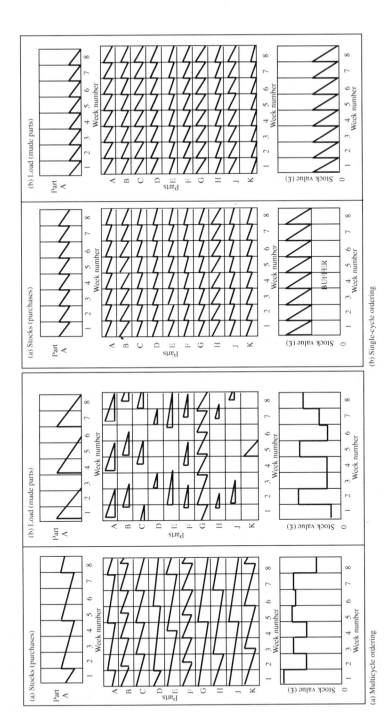

FIG. 1.8 The surge effect.

2. Parts should not be made until they are needed for assembly or sale, and shop orders should cover period needs for only a short time into the future.
3. Deliveries should not be taken from suppliers until required for assembly, or for further processing.

The main objective of this JIT approach is to reduce manufacturing throughput times and to reduce stocks. Two such systems have already been illustrated: Kanban (Fig. 1.5) and PBC (Fig. 1.4). These JIT systems are always single-cycle systems which base production on explosion from a series of short-term production programmes, or in other words on *flexible programming*.

Just-in-time systems introduce two new problems to production control. First, they usually require some reduction in throughput times (manufacturing lead times), and, secondly, they require some reduction in setting-up times. The first problem is normally solved by using Group Technology GT and by close-scheduling, or starting following operations on the parts in a batch, before all the parts have been completed at the previous operation. As the Japanese have demonstrated, set-up-time reduction is not difficult if the effort is made to achieve it. Some possible methods are examined in Chapter 9.

1.6 The economic-batch quantity

The main reason why the West has been unable to compete in manufacturing with Japan is that with our traditional methods of PC we are unable to achieve their high rates of stock turnover. This in turn is mainly due to the long-held acceptance of the economic-batch quantity (EBQ) theorem in the West. (This is called the economic lot size in the USA.)

The EBQ theorem was first published by the late F.W. Harris in Chicago in 1913. He said, as is illustrated in Fig. 1.9, that for any batch of parts, 'there is a fixed preparation cost per batch. The preparation cost per part falls exponentially as the batch quantities are increased, and as this fixed preparation cost is spread over more and more parts.' At the same time, the increase in batch quantities has a linear relationship with both the investment in stocks and with the cost of holding stocks. The combined effect of these two types of change (he believed) produces a total-cost curve with a pronounced minimum cost at one particular batch quantity, known as the economic batch quantity (EBQ) (or the economic lot size in the USA). He believed that, for minimum production cost, all items produced in factories should be made in their own calculated, special, EBQ quantities. This theorem is false for five main reasons.

First, it is a typical example of suboptimization which breaks the gestalt law attributed to Aristotle. The choice of hundreds of different optimum values (one for each part,) which ignores the effect of all these choices on the total system cannot possibly find a true optimum for the system as a whole.

Second, the use of EBQs produces a multicycle system, with all their deficiencies of ordering based on unreliable long-term forecasts, of the surge effect, and of complexity.

AN OUTLINE OF PRODUCTION CONTROL

The economic batch quantity (EBQ) theorem states: for each part in production there is one easily calculated batch quantity which must be used to obtain minimum production cost.

This theorem is false because:

1. It breaks the Gestalt law (attributed to Aristotle), which states (in current terminology): many suboptimums never find a total optimum.
2. It is a multicycle system, and it is therefore unreliable due to the surge effect.
3. Changes in batch quantity (BQ) have very little effect on the unit cost (UC), which is mainly fixed in relation to the BQ. The traditional model exaggerates the value of the EBQ.

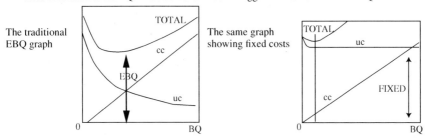

The traditional EBQ graph

The same graph showing fixed costs

Key: UC = unit cost (Preparation sum of operation + costs) and CC = carrying cost.

4. There are four main batch quantity variants. They are independent variables and each has a different economic effect. The EBQ theory treats all these variants as one single variable.

Reduce		Cost	Investment	Return/investment
Order quantity	OQ	↑ More orders	–	↓
Run quantity	RQ	↓ Stock-holding costs	↓ Stocks	↑
Set-up quantity	SQ	↑ Set-up cost per part	–	↓
Transfer quantity Process organization	TQ(1)	↑ Materials handling cost–more handling moves	–	↓
Ditto with GT	TQ(2)	–	–	–

5. It is always more economical to invest in the reduction of set-up times and run quantities than to invest in stocks by using large run quantities to spread the cost of setting up over many parts.

FIG. 1.9 The case against the EBQ theorem.

Third, the EBQ theory, as illustrated in Fig. 1.9, gives a misleading idea of the relationship between batch quantity and cost. Most costs have only a small variation with changes in the batch quantity; they are mainly fixed costs in relation to batch-quantity changes, for example, rent, rates, salaries, heating, lighting. The EBQ theorem gives a highly exaggerated view of the relationship between the batch quantity and cost. The diagram showing the fixed costs in Fig. 1.9 gives a truer picture of the relationship which would exist if the late F.W. Harris's assumptions about the batch quantity were correct.

The fourth deficiency of the EBQ theory is that it is based on the use of an extremely inefficient concept of the batch quantity. There are many ways in which parts can be combined into batches for convenience in production. The four most important ways for PC are:

(1) *the order quantity* (OQ) or the number of parts authorized for production by an order;
(2) *the run quantity* (RQ), or the quantity of a particular part produced on a machine before changing to make some other part;
(3) *the set-up quantity* (SQ), or the quantity of parts, not necessarily all the same, which are produced on a machine before changing the tooling set-up;
(4) *the transfer quantity* (TQ), or the quantity of parts transferred as a batch between two machines or other work centres which carry out successive operations on the part.

All four of these different batch-quantity types are parameters, or variables to which a manager can assign arbitrary values at will. All four are independent variables. The value of any one of them can be changed (within a restricted range) without changing the values of the other three. As illustrated in Fig. 1.9, changes in each of these parameters have a different effect on the stock investment and on costs. Reducing the order quantity increases reducing the OQ increases both the number of orders per year and the ordering costs; Reducing the run quantity (RQ) reduces stocks (see Fig. 1.10) and it also reduces stock-holding costs, reducing set-up quantity (SQ), and increasing the number of different set-ups increases SU cost per part, and reducing the transfer quantity (TQ) increases the materials-handling costs (more handling moves) with process organization, but has no significant effect with GT, because machines are close together. These relationships are reversible.

It is highly desirable that the different parameters should retain their independence, so that, one can for example, increase set-up quantities and reduce run quantities at the same time. In EBQ theory, all these four types of batch quantity are treated as a single parameter called the *batch quantity* (or lot size), and they are changed as a single unit. This makes it impossible to achieve the advantages which are obtainable when the four constituent parameters retain their independence.

AN OUTLINE OF PRODUCTION CONTROL

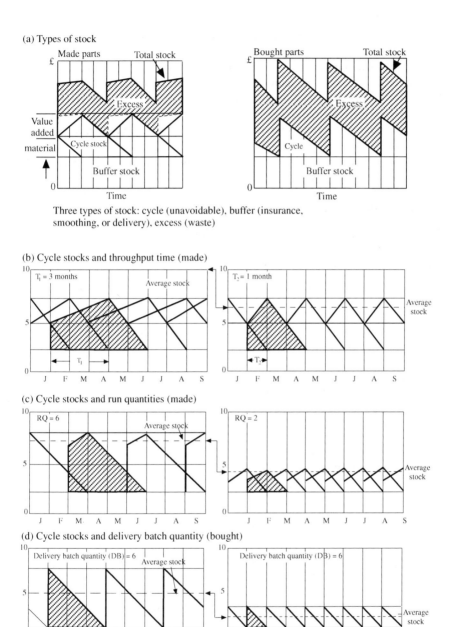

FIG. 1.10 The effect on stocks of changes in the throughput time, the run quantity, and the delivery batch quantity.

Fifth and finally, EBQ theory treats the set-up time as a constant, and it is reasoned that the cost per piece for parts can only be reduced by manufacture in large batches (investing in stocks) in order to spread the cost of setting up. The Japanese have shown that the set-up time and cost are not invariable constants. They can easily be reduced at a relatively low cost. The EBQ theorem is nonsense, because it is much more profitable in practice to invest in set-up-time reduction, so that parts can be made economically in small run quantities, than it is to invest in stocks by manufacturing in large run quantities with a view to spreading existing set-up costs over many parts.

The EBQ theory is pseudoscientific nonsense. Its total rejection is essential for the future development of PC.

1.7 Dispatching

The third level of PC, known as dispatching, deals with the methods used to manage the work of completing products or components in company workshops. In component-processing departments, it covers such tasks as:

(1) operation scheduling;

(2) inspection;

(3) setting-up machines;

(4) tool storage and make-ready;

(5) tool presetting;

(6) preventative maintenance for machines;

(7) tool maintenance;

(8) materials handling;

(9) good housekeeping;

(10) the removal of swarf;

(11) the maintenance of records.

In factories with traditional process organization, most of these tasks are centralized at departmental level. With GT, it is generally possible and advantageous to delegate most of them to the groups.

In assembly departments and assembly groups, many of the same dispatching tasks are needed, but there is also one new task (namely *kitting out*, or the collecting together of the sets of parts for different assemblies) which is special to assembly.

The most difficult dispatching task is usually operation scheduling. Three main methods are used in factories with process organization:

(1) the due-date-filing method;

(2) scheduling future production on Gantt charts (or an planning boards or on the computer);

(3) next-job decision rules.

AN OUTLINE OF PRODUCTION CONTROL

The *due-date-filing method* is illustrated in Fig. 1.11. New orders are filed on arrival in the section of the order file reserved for the machine used for their first operation. As each operation is completed, the order form is moved to the file section for the machine used for the next operation. A daily plan is produced from the files showing the next two or three jobs on each machine. A copy of each order

(a) On receipt of an order

1. The order is filed in section for machine for first operation in order file.
2. A due-date copy of the order is filed by the due date in Due-date file.
3. Job cards are filed by the part number and by the operation number in job card file.

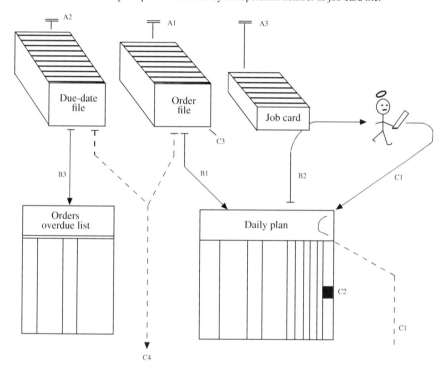

(b) Daily

1. The order file shows which jobs go on the daily plan.
2. The daily plan shows which job to issue to which worker.
3. The orders overdue list shows which jobs are overdue.

(c) On completion of an operation

1. The worker returns the job card when the job is finished. It goes to the wages department.
2. The dispatcher brings the daily plan up to date.
3. The order is moved to the section for the machine for the next operation. (See A1).
4. After the last operation, the order and due-date copy are sent to the PC office.

FIG. 1.11 The due-date-filing dispatching system.

is also filed in a separate file by due date. This is used to produce a periodic list of overdue orders. This due-date-filing method is simple and reliable in use, but it has one important deficiency; namely, that all the parts in each batch must be completed at one operation, before the next operation can start. In other words, close-scheduling cannot be used, and the throughput times are high.

The *scheduling method* based on Gantt charts or planning boards is illustrated in Fig. 1.12. This method, which dates back to the work of an American

(a) *On receipt of an order*

1. The order is scheduled on a Gantt chart or planning board.
2. The order is then filed by the due date in the order file.
3. Job cards are filed by the part number and by the operation number.

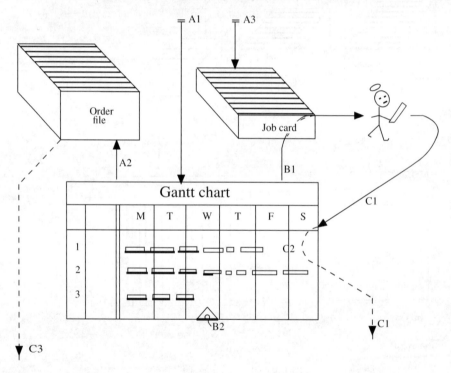

(b) *At all times*

1. The Gantt chart shows which job to issue to which worker.
2. The Gantt chart shows which jobs are overdue.

(c) *On completion of an operation*

1. The worker returns the job card, which goes to the wages department.
2. The dispatcher brings the Gantt chart up to date.
3. After the last operation, the order is sent to the PC office.

FIG. 1.12 The Gantt-chart system of dispatching.

named Gantt in the early 1900s, has the advantage that close-scheduling (or starting following operations before the preceding operations are finished on all the parts in a batch) is possible. It also however, has the serious deficiency that crises on one machine (such as absentee workers, scrap, material shortages and machine breakdowns) affect large parts of the schedule, and generally mean that it must be redrawn. The OPT method uses computer operation scheduling planned a week at a time, but in practice its schedules seldom remain intact for the whole week. The success of the OPT method has depended more on Dr Eli Goldratt's principles than on its computer software for scheduling.

Next-job decision rules is the third method of operation scheduling. It is based on the idea that rules can be established for answering the question, What should be the next part to be machined at this work centre?, and that these rules can be applied progressively as the need arises. Scheduling progressively in this way should provide the necessary flexibility to give quick decisions when there are crises. In practice, this approach has not been completely successful. It fails, for example, to give the notice needed for the collection of tools for make-ready for setting up, and for presetting.

The combination of GT plus PBC introduces the possibility of a fourth method of operation scheduling. In many industries if there is sufficient capacity it is possible with PBC to preplan the sequence for loading first-operation machines and then to rely on queuing discipline, with perhaps some additional sequencing-priority rules for special cases to provide a flexible schedule which will complete the period load by the end of each period. This method is known as *launch-sequence scheduling* (LSS); it is described in greater detail in Chapter 11.

1.8 Summary

Currently, the most widely used PC methods are stock control (SC) and MRP. These methods are obsolete in their traditional form, because they cannot achieve the high rates of stock turnover needed to compete with the Japanese. It has been shown that this failure is due to: the basing of ordering on long-term forecasts, which are always unreliable; the use of multicycle ordering with its unavoidable surge effect; to the inherent complexity of multicycle systems; and, particularly in the case of stock control, the magnification of the demand-variation effect (industrial dynamics).

Period batch control, if used with GT, overcomes these deficiences. It bases ordering on a series of short-term programmes, which are much more reliable than long term programmes, and because it is a single-cycle ordering system, it is not susceptible to the surge effect or to the full magnification of the demand-variation effect, and it is therefore a comparatively stable system. Finally, PBC operates efficiently with a very small investment in stocks, reducing both the investment and the stock-holding costs, to give a geared increase in the rate of return on investment.

2
THE MATERIAL-FLOW SYSTEM AND GROUP TECHNOLOGY

2.1 Introduction

One definition of production control (PC) states that it is the function of management which regulates the flow of materials through the materials-flow system (MFS) of an enterprise. The MFS is the system of all the routes in a factory along which materials flow between the places where work is done on them. An MFS can be described in simple terms as a road map, or as a route map for the flow of materials.

The MFS is a static system of routes. It can vary between a simple unidirectional-flow system and a system of extreme complexity. When materials are made to flow through the MFS, the result is a dynamic flow system of a much higher order of complexity than the MFS. From experience, these dynamic systems are easier to regulate and control if they are based on a simple MFS rather than on a complex MFS.

The complexity of the MFS depends on the way in which the work is organized in the factory. There are two main methods of organization:

(1) *process (or horizontal, or functional) organization*, in which organizational units specialize in different processes.

(2) *product (or vertical) organization*, in which organizational units complete particular products or parts, or sets (families) of products or parts.

Figure 2.1 shows the effect of these different types of organization on the MFS. Because different parts require different combinations of processes in different sequences, process organization always produces a very complicated MFS. In other words, companies with process organization tend to suffer from the 'drunken spider,' or the 'spaghetti' syndrome.

The simplest form of product organization is that which is based on continuous line flow (CLF). The machine and other work centres used to make a product or part are laid out in the sequence in which they are used so that there can be a continuous flow of materials between them. The use of CLF is limited to cases where all, or most, of the items produced use the same processes in the same sequence. It is used in most process industries (cement, sugar, cheese, etc.), and it is also used, to a limited extent, for the production of more complex items which are required in very large quantities (mass production). It is a feature of CLF systems that for maximum efficiency the

THE MATERIAL-FLOW SYSTEM AND GROUP TECHNOLOGY

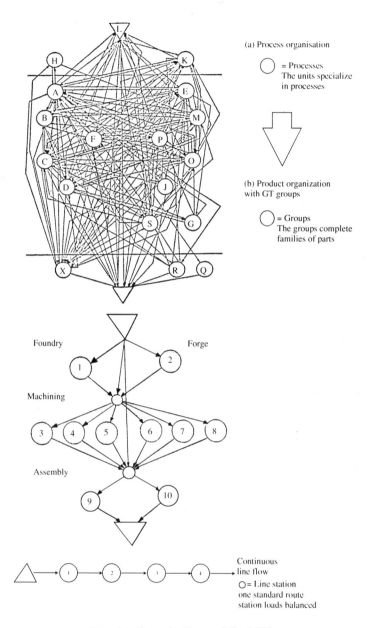

FIG. 2.1 Organization and the MFS.

loads in machine-hours or man-hours at each station, must be the same, giving a balanced line.

Group Technology (GT) is a more recent form of product organization in which there is a total division of machines into Groups which complete all the components they make. It is not subject to the same limitation as CLF (that all parts must use the same processes in the same sequence). With GT groups, different parts may use different combinations of the machines in a group, in different sequences. Group Technology is mainly used in jobbing and batch-production factories.

The choice of the PC system for an enterprise is profoundly affected by the type of MFS used. In particular, there are types of industry where Period Batch Control (PBC) is only possible if it is used in conjunction with GT. We are obliged, therefore, in this study of PBC, to examine the nature of GT at an early stage.

2.2 Elements of GT

The term *Group Technology* was first used by Professor Mitrovanof of Leningrad University as the title for his research into the relationship between component-shape and processing methods. Among other findings, he showed that lathes could be set up to make a number of different but similar parts, one after the other at the same set-up. It was then found that by adding other machines, groups could be formed which completed parts. Further work—notably by Graham Edwards at UMIST in Manchester, by J. Gombinski of E. G. Brisch and Partners, by G. Ranson of Serck-Audco Ltd, and by C. Allen

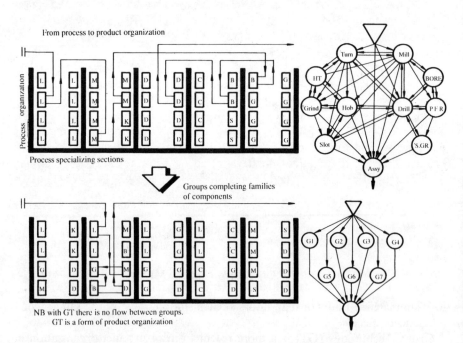

FIG. 2.2 The difference between process organisation and GT.

of Ferrantis Ltd in Edinburgh—showed that the same principle could be applied to other types of machine tool in machine shops, and to other industries such as foundries, forges, welding, wood-working, and sheet-metal workshops.

Group Technology evolved in this way into a form of product organization which can be used for component processing in any factory, with more than, say, twenty workers, where CLF is inappropriate. The main difference between a machine shop with process organization and one with GT is illustrated in Fig. 2.2.

With process organization, parts move between operations from one process-specializing section under one foreman to other sections under other foremen. At each move, they join a queue and must wait their turn for processing. The throughput times are therefore very long. With GT, the work on each component is all managed by one foreman, who can schedule the operations for much shorter throughput times, with much less stock. An example of such a group is shown in Fig. 2.3. The foreman of this group was responsible for the quality of the work produced, for those elements of the cost of the work which he could control, and for the completion of jobs by their scheduled due dates.

The group in Fig. 2.3 was installed over twenty-five years ago. The machines which were used were all manually operated. Today, this group contains three computer-controlled CNC turning centres. One of these has the facility to do second-operation work, and the two second-operation lathes have been eliminated. A new machining centre has replaced the two mills and the two drills in the original group. This group provides a useful example of the evolutionary route towards automation. The original group, like most GT groups, was a flexible manufacturing system (FMS) with a great deal of manual intervention. The latest version is not yet fully automated, but it has much less human intervention.

It is interesting to note that the division into groups in a company, once found, tends to be stable. The two oldest examples of GT in Great Britain are in the factories of Serck-Audco Ltd at Newport, Staffordshire—making valves for the oil industry—and of Ferrantis Ltd in Edinburgh—making radar equipment. Apart from one additional group in each, they have the same groups today as they had when they started with GT twenty-five years ago. The design of the products has changed enormously. The old manually operated machines have been replaced by CNC, or DNC, controlled machines, but both companies still make the same general types of product, and they still require their own mixture of, say, small rotational parts, box-like cases, gears, shafts, bushes, and/or others. They still require the same division into groups.

2.3 GT stages

With the simpler types of industry, such as the process, implosive, and square industries (see Fig. 1.3), groups normally complete products at a single stage (as illustrated in Fig. 2.4(a)).

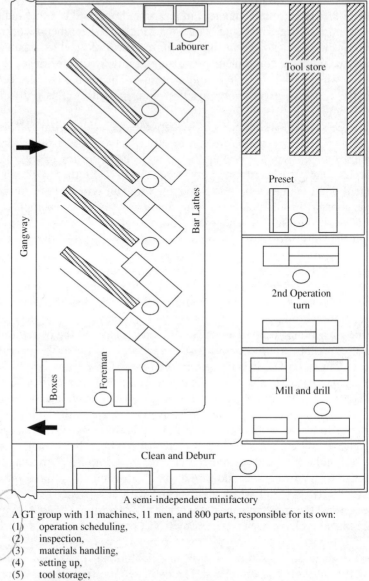

A semi-independent minifactory

A GT group with 11 machines, 11 men, and 800 parts, responsible for its own:
(1) operation scheduling,
(2) inspection,
(3) materials handling,
(4) setting up,
(5) tool storage,
(6) minor tool maintenance,
(7) minor preventative-maintenance inspection,
(8) housekeeping.

FIG. 2.3 A GT group.

With the more complex 'explosive' industrial systems, groups may be needed at a succession of stages, in order to complete products. For example, if a

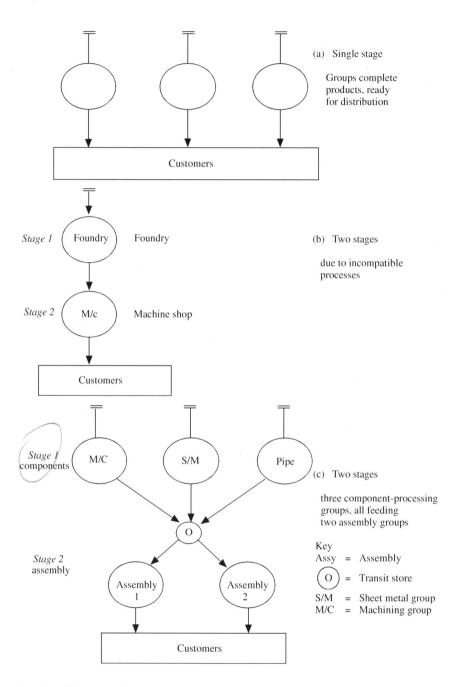

FIG. 2.4 GT processing stages. P.c. 'oh', transit store; s/m, sheet-metal group; m/c, machining group; assy, assembly.

company makes sand castings in a foundry which are then machined in a precision machine shop, it may be impractical, due to process incompatibility, to form a group which does both processes, and groups must be formed at successive stages (see Fig. 2.4(b)).

Again, if the same types of component are needed in several different assembly groups, it may be impractical to make the same types of component specially for each group. Components will then have to be made in component-processing groups at one stage, for distribution to assembly groups at a second stage (see Fig. 2.4(c)).

It is desirable that the number of stages should be kept as small as possible in order to minimize the throughput times and the stock investment. In a later chapter (Chapter 8), examples are given of processes which were traditionally seen as incompatible, and which can now be found in the same groups.

---

1. **Short throughput times**
   Because machines are *close together* under one foreman, and groups complete all the parts they make. This also gives: -

   (a) low stocks,
   (b) low stock-holding costs,
   (c) better customer service.

2. **Better quality** (fewer rejects)
   Because groups *complete parts* and machines are *close together under one foreman*.

3. **Lower materials-handling costs**
   Because machines are *close together under one foreman* and groups complete parts.

4. **Better accountability**
   Because groups *complete parts*. The foreman can be made responsible for the *cost*, the *quality*, and completion by the *due date*, giving:

   (a) reduced indirect labour costs,
   (b) more reliable production.

5. **Training for promotion**
   Process organization only produces specialists. Group Technology provides a line of succession because a group is a minidepartment.

6. **Automation**
   Group Technology is the first evolutionary step in automation. A group today is an FMS with some manual operations.

7. **Reduced set-up time**
   Similar parts are brought together on the same machines.

8. **Morale and job satisfaction**
   Most workers prefer to work in groups, where they can influence the way the work is done.

FIG. 2.5 advantages of GT.

2.4 The advantages of GT

Figure 2.5 lists the main advantages of GT over the traditional methods of process organization. It can be seen that these advantages arise because the machines in a group are close together under one foreman, and because groups complete all the parts they make. These two conditions provide the essential objectives for the design of a successful GT system. It will be realized that most of the advantages listed in Fig. 2.5 are not gained if groups fail to complete the parts they make and if there is back-flow or cross-flow between the groups.

In effect, two different types of production change are found in manufacturing. First, there are the *component changes* in design or processing methods which are introduced to improve the quality or to reduce the cost of particular parts. Secondly, there are the *total-system changes*, such as the change to GT, which affect most of the parts and achieve a major change in the system as a whole.

2.5 Planning the division into groups and families

Most of the original applications of GT were planned using *classification and coding* (C&C) of the component drawings, based on component shape and function. Today, C&C has been largely superceded by *production-flow analysis* (PFA), which finds the division into groups and families by analysing the information contained in component-route cards.

Classification and coding was originally chosen to find the families for GT on the assumption that: parts with the same shape or function can all be produced by the same group of machines. Although this is true for the majority of parts, there are too many exceptions for reliability. Parts, for example, may be similar in shape or function, but they may need to be made on different sets of machines because they differ in size, material, requirement quantities, or in manufacturing tolerances.

Other problems with C&C are that: it fails to find the many parts in any factory which differ greatly in shape or function from those in a particular C&C group, but which ought to be made in that group because they can only be made on that set of machines; or because they require the use of one or more machines from that set of which there is only one in the factory. It only finds families of parts; it does not help in finding the associated groups of machines; and it requires an expensive special C&C database compared with the readily available route-card database—already provided for PC which is used with PFA.

Classification and coding has advantages for other purposes in its own right, but even if there is already such a C&C system in a company, it would be ill-advised to use it to plan GT because it will not find a total division into groups and families.

The best technique for planning GT is 'Production-Flow Analysis', or PFA. This is a set of subtechniques used one after the other to progressively simplify

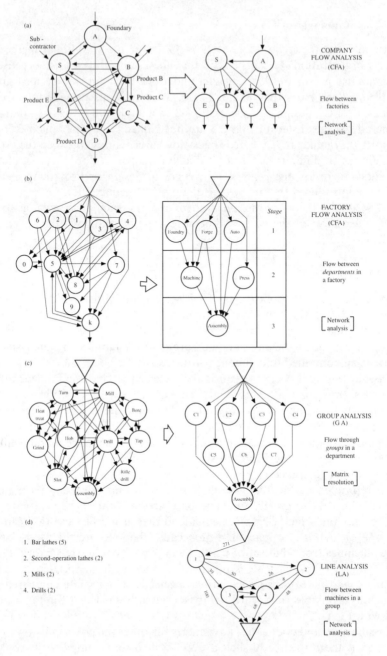

FIG. 2.6 Subtechniques of PFA.

the MFS of the enterprise (see Fig. 2.6). First, company-flow analysis (CFA) simplifies the flow between different factories between or the major divisions in

THE MATERIAL-FLOW SYSTEM AND GROUP TECHNOLOGY   31

a large company. Next, factory-flow analysis (FFA) divides, in turn, each factory (or division) into departments which complete all the parts they make and which are connected by a simple unidirectional 'material flow system.' Group analysis (GA) is then used to divide, in turn, each department into groups and families. If a department completes all the parts it makes and if it employs, say twenty workers or more, it is always possible with GA to divide it into smaller groups, which again complete all the parts they make. Next, line analysis (LA) analyses the way in which materials flow between the machines in each group, to provide the information needed for plant layout. Finally tooling analysis (TA)—which is not included in Fig. 2.6—considers each machine, in each group in turn, and analyses the tools it uses to find *tooling families* of parts which can be made at the same set-up, to find the optimum loading sequence, and to rationalize the tooling. Tooling analysis, like GA, is based on matrix resolution, whereas the other three subtechniques of PFA use network analysis.

2.6 FFA and GA

The two subtechniques of production-flow analysis (PFA), which find groups and families in a factory, are factory-flow analysis (FFA) and group analysis (GA).

Factory flow analysis analyses the flow between major processes to find a division into departments which complete all the parts they make. In practice, the departments based on strict process specialization are similar to those found by FFA. Departments such as machine shops, cold-forging and welding departments, and sheet-metal shops do normally complete most the parts they make. The exceptional routes, where parts have operations in more than one component-processing department can be found by FFA, which eliminates these divisions by the redeployment of a few machines.

In one cold-forging and welding department, for example, FFA revealed that some parts they made were routed to a machine shop for intermediate milling (weld-edge preparation) and drilling operations. The moving of a universal mill and a radial drill into the cold-forge department made it possible for the department to complete all the parts it made. Figure 2.7. shows the MFSs before and after FFA in that factory.

If the departments in a factory complete all the parts they make, and if they are large enough (employing, say, twenty workers or more), then it is generally possible to divide them into groups which complete all the parts they make.

The process of GA is equivalent to the *resolution of a matrix* containing machine tools and made parts (see the diagrams in Fig. 2.8). Because the machine/part matrix in most companies reflects their use of process organization, with departments which do not complete all the parts they make, and because most departments in industry make hundreds of different parts,

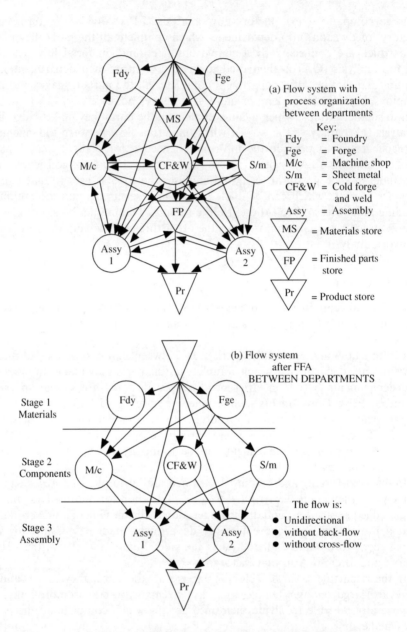

FIG. 2.7 Materials flow before and after FFA.

making very large matrices, GA cannot be based on the existing part/machine matrix. It has to be modified to form minimatrices called *modules*. Each module is based on a key machine taken from the plant list, and shows all the

(a) Component machine chart

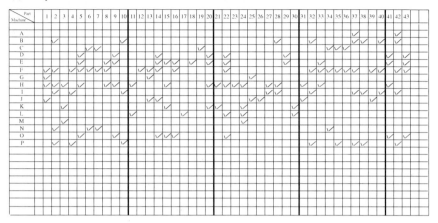

(b) Division into groups and families

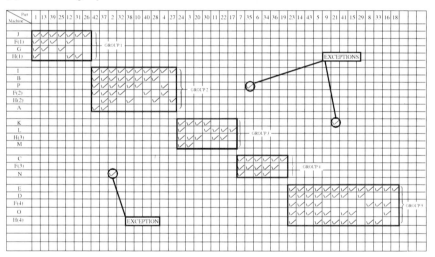

FIG. 2.8 Group analysis: the resolution of a matrix.

remaining parts which have not already been included in previous modules which have operations on the key machine.

The division into modules also makes it possible to give precedence to those machines for which there is only one, or very few, of that type, and to those which do work which it would be impossible to transfer to any other machine in the factory. The selection of the key machines from the plant list is based on a special SICGE code. Machines are *special* (S) if there is only one of them and if it would be extremely difficult to transfer any operation done on that machine to any other machine. They are *intermediate* (I) if they have similar characteristics to an S machine, but there is more than one of the type. They are common (C) if there is a high possibility that operations on one of the machines can be easily transferred to some other machine. Examples include: most lathes, mills, and drills in machine shops. Most of the machines in most factories are of the C class. Next, there are G machines which are used for a wide variety of different parts and cannot conveniently be included in groups. They are normally required at stage boundaries, and they will often be accommodated as a new stage. Painting and electroplating plant in explosive industries provide examples. Finally, the category E (equipment) covers items used to support manual operations. Benches, vices, centres for rotating shafts, polishing buffs, and hand-held power tools are examples.

The key machines used to form modules are selected in turn from a *special plant list* (SPL), in which they are listed in SICGE sequence, and in which inside each category they are listed in the sequence of the increasing number of parts F, with operations on each machine.

The most common question asked by engineers when they first hear of GT is, 'If I have two groups, both of which need the services of one special machine, how do I form groups which complete all the parts they make?'. The answer is that this cannot happen with PFA. There is only one of each S-machine, so it can be and is, only installed in one group. All parts with operations on that machine are made in the group where it is installed.

When the modules have been formed and have been combined to form groups, there may still be ten per cent or more of the parts in a group for which the route cards show operations routed to machines (generally, C class machines) which have not been allocated to the group. At this stage, the production engineers re-allocate such operations to other C class machines which are in the group. It is the high level of routing flexibility in industry which makes it unlikely that a case will ever be found where GT is not possible.

2.7 Assembly groups

The term *Group Technology* was originally reserved for component-processing groups. Very similar groups are also found which complete assemblies or major stages in assembly. These assembly groups had a different history, and as they are easier to plan they do not need the services of a special technique such as GA, although FFA is still used to plan departments and stages. Figure 2.9 shows the five main types of assembly group that are found in practice.

Assembly groups of the monogroup type have been used in small companies for many years, but their introduction in their modern form can be traced, among other sources, to research at the Tavistock Institute of Human Relations, which indicated that workers had greater job satisfaction and were more efficient if they worked in groups (teams) which completed products than if they worked as isolated process specialists.

THE MATERIAL-FLOW SYSTEM AND GROUP TECHNOLOGY

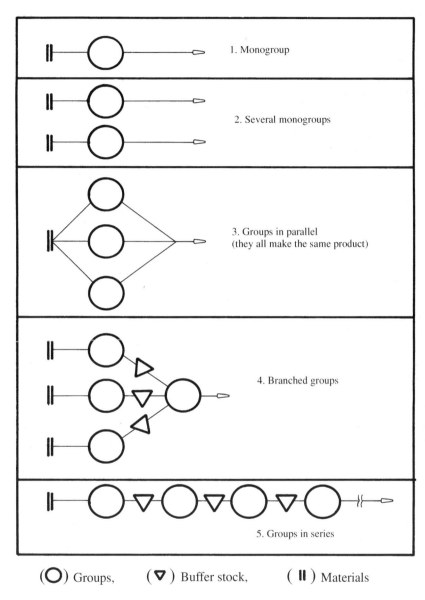

(◯) Groups,   (▽) Buffer stock,   (‖) Materials

FIG. 2.9 Organization of assembly groups.

At the end of the Second World War, the Norwegian government was faced with the task of rebuilding its industry. In collaboration with the Norwegian Employers Federation and the Norwegian Trade Unions, they agreed that group-production methods were desirable. They employed two behavioural scientists from the Tavistock Institute (Trist and Emery) and one from Norway (Thorsrud) to introduce groups in Norwegian companies. Together they introduced sample groups in over a hundred companies before the experiment was abandoned.

The Norwegians did not themselves gain greatly from their efforts, mainly because they never achieved a total change to GT in any single company, but their work was widely reported and it had a major effect in several other countries. In Sweden, for example, in the affluent 1960s, the Volvo car company had difficulty in recruiting Swedes to work on their machine-paced, conveyorized, car-assembly lines. Finnish and other workers from poorer countries, attracted by the high wages, went to Sweden to take these jobs. Gullenhammer, the President of Volvo, believed there was no future for a country which had to import workers to do work which its own nationals would not accept. He told his production engineers to design a new assembly factory where Swedes would be willing to work, pointing to the Norwegian experience as a possible approach. The result was the very successful Volvo group-assembly plant for cars at Kalmar in Sweden.

Other companies which moved at this time from long machine-paced assembly lines to groups were Phillips, in their factory making television sets at Eindhoven in Holland, and Olivetti, who made electronic calculators at Ivrea in Italy. In all these cases, it was expected that the production costs would rise, but that the change to groups would be worthwhile if it reduced the alienation of the labour force. In the event, the groups, in all cases, were also more profitable than the lines they replaced.

More recently, groups have tended to replace assembly methods based on the progressive assembly of subassemblies in batches for stock. The introduction of flow-control PC systems, led to the development of parts lists (bills of material) broken down progressively into sets of parts for major assemblies, subassemblies, sub-subassemblies, and so on. Some engineers believed that the most economical method of assembly would be to make the subassemblies at these levels in large batches for stock. This method did reduce the direct labour cost of assembly, but it induced very high administration costs, very long throughput times, heavy stocks, and high stock-holding costs. Major increases in profitability can be achieved by changing from this method to groups. Figure 2.10 shows an assembly group for a machine tool, which is designed to complete products continuously and to make all its own sub-assemblies, but only when they are needed for incorporation in the finished products.

2.8 Materials transfer between GT-component-processing and assembly groups

The main objectives of group production methods are to simplify the material flow system; to reduce stocks and stock-holding costs, and to improve accountability by making it possible to delegate the authority and responsibility for production performance to those who do the work on the factory floor.

THE MATERIAL-FLOW SYSTEM AND GROUP TECHNOLOGY 37

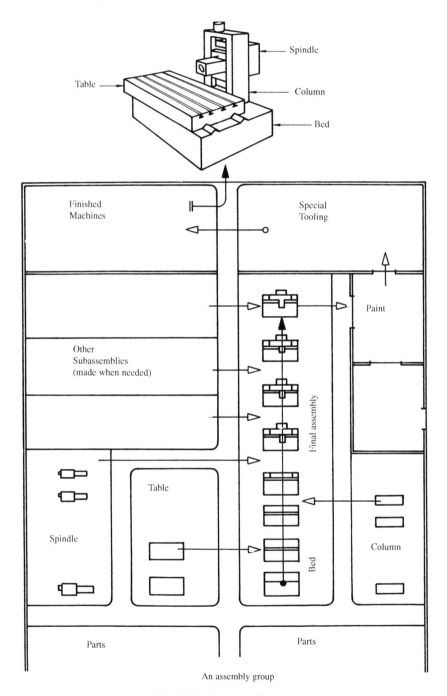

FIG. 2.10 An assembly group.

Because, with single-cycle ordering systems such as PBC, component-processing groups make parts each period in the quantities needed for assembly in the following period, most of the problems of kitting-out for assembly are eliminated. The transfer of materials between component processing and assembly may still be complex, however, if the company makes a wide range of products with intermittent demand, and if it is decided, for any reason, to hold stocks of some of the products or parts.

In the first case, the same assembly groups may be used to assemble different products in each period. It will save confusion if the component parts for each product included in each period production programme are collected outside the groups in a *transit store* and if they are then moved together in sets to the appropriate assembly groups at the beginning of each new period. In the second case, the long-term objective may be to achieve zero stock, but it may be necessary to hold some stock of some finished products if sales are only possible from stock. A small fluctuating *smoothing* stock may also be legitimate, if it saves capacity which would otherwise be lost. It may be cheaper to hold some smoothing stock than to build a bigger factory. Again, in the case of components, it may be necessary to hold some stock of parts for sale as

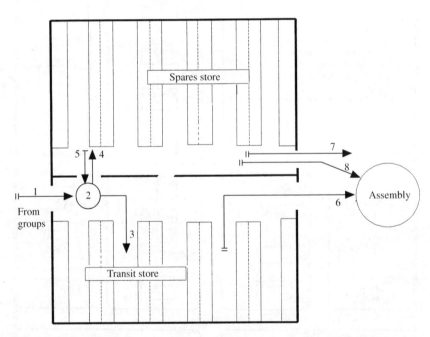

FIG. 2.11 A transit store: (1) receipts from groups, (2) quantity stock, (3) the accumulation of sets of parts for assembly (4) the sending of surplus parts to the spares store, (5) the make-up of assembly shortage from the buffer, (6) the issuing of sets to assembly at the end of the period, (7) the sale of spare parts, and (8) the replacement of assembly scrap from the buffer.

spare parts, and in the early stages of introduction it may also be desirable to carry some buffer stock as an insurance against errors in supply.

Methods for controlling product-stock levels will be examined in a later chapter. A method for handling component transfers between compont processing and assembly is illustrated in Fig. 2.11. In this case, all the parts received from component-processing groups and in period sets on call-off from suppliers are routed to a transit store, where they are counted and checked for quantity against copies of the group-list orders and purchasing-call-off notes.

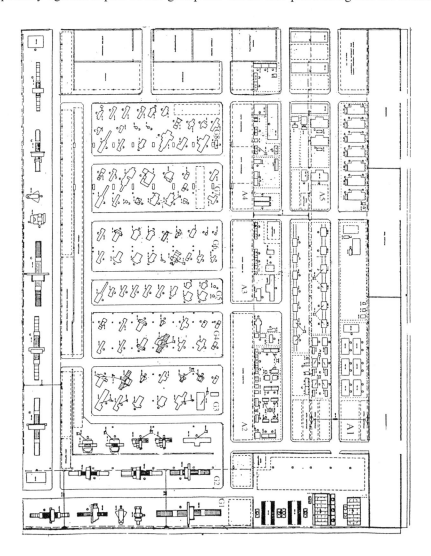

FIG. 2.12 Total GT in a factory making grinding machines.

The exact quantities needed for assembly are stored on portable racks, which are moved at the end of the period into appropriate positions in the assembly groups. Any surplus is stored in the spares store, and is used to meet both spares sale demand and any demand for scrap or shortage relacements from buffer stocks. The method used to control these stocks is again described in a later chapter.

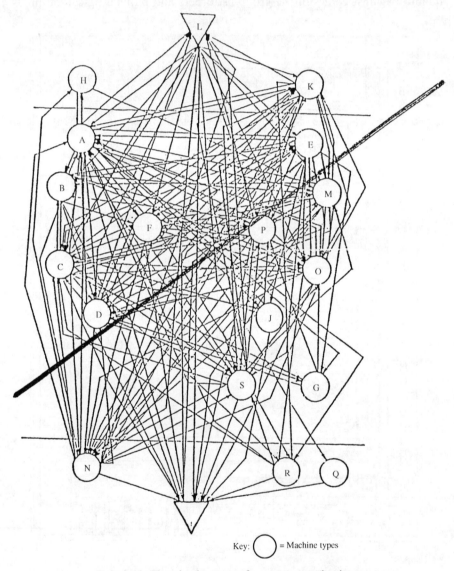

FIG. 2.13 The obsolescence of process organization.

THE MATERIAL-FLOW SYSTEM AND GROUP TECHNOLOGY   41

2.9 Total Group Technology

The final objective with GT should be a total division into groups. Figure 2.12 shows such a division in a factory making a range of grinding machines. It can be seen that the group 1 machines for heavy machined parts are laid out in an overhead-crane-served line, with a right-angle bend in it. This is not the ideal arrangement, but it was chosen to save making new foundations and pits for the machines. It is generally possible to substitute GT for process organization in any manufacturing company.

2.10 Summary

Group Technology is a method of factory organization of the product organization type, in which organizational units (groups) complete all the parts in the sets of parts (families) which they make, and there is a total division of all the machines and other processing facilities into groups, and of all the made parts into associated families.

The early GT applications were planned using classification and coding (C&C) of component designs. This method is now obsolete for planning GT, because it does not find a total division into groups which complete all the parts they make. It has been replaced by PFA which finds groups and families by analysing the information in component *process routes*. Group-production methods can also be used for assembly, where they are replacing both long machine-paced, conveyorized, assembly lines, and the progressive manufacture of subassemblies in batches for stock.

Apart from the process industries and a few other cases where continuous line flow is possible, and apart also from small-scale industries, which are in effect already groups, it is always possible to use group-production methods. They are so much more efficient than the traditional methods based on process organization that process organization can now be described as obsolete. (See Fig. 2.13).

3

THE PRODUCTION-CONTROL FEEDBACK CONTROLS

3.1 Introduction

Production control (PC) was defined in Chapter 1 as the function of management which regulates and controls the flow of materials through the materials-flow system (MFS) of an enterprise. The term *production control* in this statement is the name of a function and is not in itself a definition.

To fully understand the meaning of the term production control, a number of other terms need to be defined. First, production is the manufacture and distribution of goods. Production control is concerned, therefore, with materials flow both inside factories and in the distribution systems which link factories with their suppliers and customers.

The hundreds of different management tasks that have to be carried out in industry can be classified into sets of closely related tasks which constitute the functions of management. The most widely accepted division finds the following eight main functions:

(1) product design;
(2) production planning;
(3) production control;
(4) purchasing;
(5) marketing;
(6) finance;
(7) personnel;
(8) secretarial.

These eight functions provide the primary classification of the applied science of production management. Production control is the function of management which includes all the management tasks concerned with the regulation and control of materials flow between the places where work is done to convert materials into products and where they are held, from time to time, in storage.

The *processes* of management are planning, direction, and control. Planning is the process of deciding what to do in the future; direction is the process by which plans are caused to be implemented; and control is the process of management by which events are constrained to follow plans. This chapter deals with control (in this management sense) in the function of production control.

As mentioned in Chapter 1, the three main controls used in PC are: progressing, loading, and inventory control. The basic method used for all three is the same. Actual achievement is measured and is compared with plans (monitoring). Any major variances discovered are communicated to the appropriate managers, by (feedback), so that they can take remedial action. This chapter studies *progressing* and *loading*. *Inventory control* is examined in the next chapter, which deals with stocks.

3.2 Systems theory

To understand the meaning of control in its management sense, at least an elementary understanding of systems theory is needed. A system is a set of variables that are so related that a change in the value of any one of them will affect the value of at least one of the others. Figure 3.1 shows a system diagramatically in the form of a rectangular box with external connections to the the three main types of variable which it contains. These are:

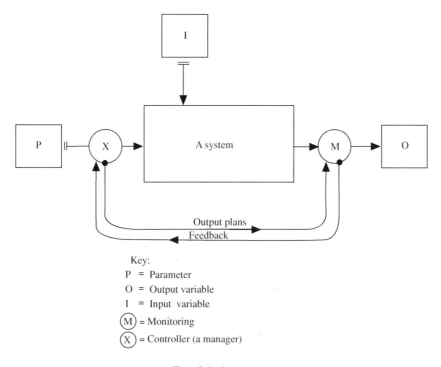

Key:
P = Parameter
O = Output variable
I = Input variable
(M) = Monitoring
(X) = Controller (a manager)

FIG. 3.1 A system.

1. Parameters, or variables to which a manager can assign arbitrary values at will. Typical examples include: selling price, run quantity, order quantity, and transfer quantity.

2. Output variables, or variables whose values can only be changed indirectly, by changing the values of appropriate parameters. Typical examples include: product output per period, throughput time, stocks, capacity, and scrap.

3. Input variables—or uncontrollable input variables—are variables which affect system performance, over which the manager has no direct control. Typical examples include: the weather, currency-exchange rates, the interest on loans, and the competition from other companies.

In their attempts to plan, to direct, and to control the operation of systems, managers start by planning future values for the output variables, planning inside the limits imposed by the processing and financial capacity of the enterprise. When making these plans, the present and forecast future values for the uncontrollable input variables are taken into account.

Managers cause these plans to be implemented (direction) by issuing programmes and orders to the factory and to it's suppliers, which are based on the planned *output-variable* values. The results achieved in meeting these programmes and orders (plans) are controlled by monitoring the actual achievement, by comparing it with the planned achievement, and by feeding back information about major variances between them to the appropriate manager so that this manager can change parameter values to steer the output values back into line with the plans.

By providing a universal model of a system, which is applicable to all types of production system, systems theory simplifies an extremely complex subject. In practice, however, there are two gaps in the theory which still have to be closed:

(1) how must the manager change the parameter values to achieve a required change in the output values?
(2) how should a production system be designed in order to maximize such desirable characteristics as reliability, flexibility, productivity, and profitability.

3.3 Connectance

The first of these problems is concerned with the way in which a change in one variable affects the values of other variables. It is concerned in other words, with the *connectance* between variables. These relationships are extremely complex. It is not only parameter changes which affect and alter the values of output variables. Changes induced in one output variable can also affect the values of other output variables. These changes are not simple limited changes. The change induced in an output variable by a parameter change may itself

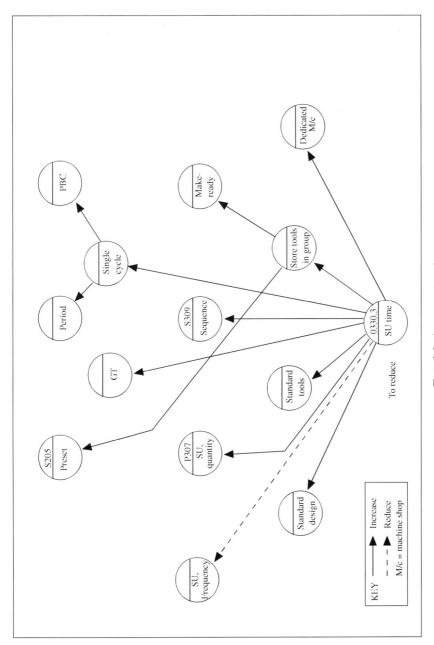

Fig. 3.2 A cure network.

induce a further change in some other variable. Each change may induce a cascade of variable changes throughout the system.

Another complication is that some variables are closely related under some conditions, for instance, one variable may have a simple linear relationship with another, so that fixing the value for one variable will also fix the value for the other. For example, with PBC and a fixed output rate, the run quantity is a simple function of the run frequency. Again, fixing the value for one parameter may limit the possible choice of value for another parameter. For example, the choice of the run quantity sets a maximum value for the choice of the transfer quantity.

When it comes to forecasting the quantitative effect of a change in one variable on another—as, for example, in the effect of reducing run quantities on the stock level—these complications make estimating very difficult, if not impossible in some cases.

In another work (J. L. Burbidge, 1984, *A production system variable connectance model*, Cranfield University, Bedford, UK), I suggested that, while it is difficult to predict the quantitative effect of one variable change on another, there is a high probability that a given direction of change in one variable will induce one particular direction of change in a related variable. For example, reducing run quantities will generally tend to reduce stocks, even though the amount of reduction is difficult to predict. It is postulated that a cybernetic approach which seeks to steer changes in output variable values in a required direction can provide managers with the power they need to regulate and control industrial performance.

Over 200 production-system variables have been studied by the author to find lists of related variables and the direction of change induced in one of them by a given direction of change in another variable under study. Figure 3.2 shows a *cure-network* diagram, produced using this database. It shows, as an example, some of the changes a manager might make if he needed to reduce the set-up time.

3.4 The design of the system

It is known that the quantitative effect of changing the value of one variable on another variable is highly dependent on the way in which the system has been designed. It can also be observed that this question of system design involves a choice between alternative solutions, or between series of alternatives. It is concerned, for example, with the choice between such alternatives as:

(1) process *or* product organization;

(2) continuous line flow *or* Group Technology (GT);

(3) single *or* multicycle ordering;

(4) batch buying *or* the call-off system;

(5) just-in-time *or* just-in-case PC.

Each of these alternatives imposes a particular strategy for parameter change. They restrict, in other words, the freedom of choice in the way parameter values are changed. There is, therefore, a close relationship between system design and connectance. Some of these choices depend on the type of product, on the type of industry in which it is made, and on the market it feeds, but other choices are more general in their application.

3.5 Progressing

Returning now to the feedback controls used in PC, the first control which will be examined is progressing. This control seeks to constrain events to follow the plans contained in programmes, orders, and operation schedules. Here we will look only at progressing for programming and ordering. Progressing for operation scheduling is examined in later chapters on dispatching and operation scheduling. (Chapters 10 and 11).

3.5.1 *Annual-production-programme controls*

Figure 3.3 illustrates a simple method of progressing, used with annual programmes in the case of PBC. It shows the weekly output and the cumulative output planned for production in a particular year.

The actual weekly and cumulative output is entered on the programme at the end of each week, and the difference (+ or −) is calculated and entered for the cumulative figures.

The data given in Fig. 3.3 provide management with some of the feedback that is required to constrain events to follow the plans made in their annual production programmes. To improve managers' understanding of the situation, however, similar programmes are also needed for planned sales deliveries to customers, and finished-product stocks, at the same weekly intervals. If any programme has to be changed, all three of the programmes will have to be altered.

A shortfall in planned sales deliveries to customers, coupled with production at or near the forecast level, and growing finished-product stocks, indicates a need for greater marketing effort. On the other hand if sales and production are both below the programme and stock is zero, this may indicate a shortage of production capacity, requiring the types of change indicated in Fig. 9.2.

3.5.2 *Short-term-production-programme control*

With short-term production programming, new short-term sales, production, and stock programmes are issued at the beginning of each period (say a week), each for completion by the end of that period. Subject to smoothing, these programmes are generally based on the actual sales orders received.

It is essential that the load in man-hours and/or machine-hours imposed by the production programme should not exceed the capacity available. There is a need with short-term programming for accurate methods for measuring the

Product A

Week number	1	2	3	4	5	6	7	8	9	10	11	12	13	14	15	16	17	18	19	20	21	22	23	24	25	26	Total
1 Annual programme	12	12	12	12	12	12	13	13	13	13	13	14	14	14	15	15	15	16	16	16	17	17	17	18	18	18	377
2 Cumulative programme	12	24	36	48	60	72	85	98	111	124	137	151	165	179	194	209	224	240	256	272	289	306	323	341	359	377	377
3 Actual production	13	10	9	14	14	9	11	16	16	12	12	10	10	15	20	20	20	20	15	15	15	15	15	16	18	20	380
4 Cumulative production	13	23	32	46	60	69	80	96	112	124	136	146	156	171	191	211	231	251	266	281	296	311	326	342	360	380	380
5 Difference (+)	1	–	–	–	–	–	–	–	–	1	–	–	1	–	–	–	2	7	11	10	9	7	5	3	1	3	
6 Difference (–)	–	1	4	2	–	3	5	2	–	–	1	5	–	8	3	–	–	–	–	–	–	–	–	–	–	–	

Week number	27	28	29	30	31	32	33	34	35	36	37	38	39	40	41	42	43	44	45	46	47	48	49	50	51	52	Total
1 Annual programme	18	18	18	18	18	18	17	17	17	17	16	16	16	16	15	15	15	14	14	13	13	12	12	12	12	12	775
2 Cumulative programme	395	413	431	449	467	484	501	518	535	552	568	584	600	616	631	646	661	675	689	702	715	727	739	751	763	775	775
3 Actual production	20	20	18	18	18	15	15	15	15	17	16	16	16	16	16	16	16	14	14	13							
4 Cumulative production	400	420	438	456	474	489	504	519	534	551	567	583	599	615	631	647	663	677	691	704							
5 Difference (+)	5	7	7	7	7	5	3	1	–	–	–	–	–	–	–	1	2	2	2	2							
6 Difference (–)	–	–	–	–	–	–	–	–	1	1	1	1	1	1	–	–	–	–	–	–							

△

FIG. 3.3 The progress record of an annual production programme. △ = Date when record last made up, at end of week 46.

THE PRODUCTION-CONTROL FEEDBACK CONTROLS 49

Week number 21			Progress record				
Product			Programme	Actual	Difference		Action
Type	Variant	Name			+	−	
A	1 2 3 4		22 4 6	22 4 6	− − −	− − −	
Total			32	32	−	−	
B	1 2						
Total							
C	1 2 3		 51 	 48 		 3 	Make 3 on overtime
Total			51	48	−	3	
D	1 2 3 4 5						
Total							
E	1						
F	1 2						
Total							
G	1 2 3						
Total							
Grand total			83	80	−	3	

FIG. 3.4 The progress record for a short-term programme.

load and the capacity. This is easy to arrange in the case of assembly industries, but it is more difficult for those industries where the products are components.

Figure 3.4 shows a progress record for short-term programming, in this case for a company making a range of different products, not all of which are ordered every period. The most important entry on this form is the minus 3 value for variant number 2 of type C.

3.5.3 *Reconciliation of annual and short-term programmes*

One other task which can be seen as a part of progressing is the reconciliation of annual and short-term-production programmes. The short-term sales

Week number	1	2	3	4	5	6	7	8	9	10	11	12	13	14	15	16	17	18	19	20	21	22	23	24	25	26	Total
Annual programme	12	12	12	12	12	12	13	13																			
Cumulative annual programme	12	24	36	48	60	72	85	98																			
Flexible programme	13	10	9	10	10	9	10	11																			
Cumulative flexible programme	13	23	32	42	52	61	71	82																			
Difference (+)	1	–	–	–	–	–	–	–																			
Shortage on annual number programme (–)	–	1	4	6	8	11	15	16																			
Week number	27	28	29	30	31	32	33	34	35	36	37	38	39	40	41	42	43	44	45	46	47	48	49	50	51	52	Total
Annual programme																											
Cumulative annual programme																											
Flexible programme																											
Cumulative flexible programme																											
Difference (+)																											
Shortage on annual programme (–)																											

FIG. 3.5 The reconcilliation of annual and short-term programmes.

programme is based on sales orders received per period. The short-term-programme production programme is found by smoothing this sales programme. The annual production programme is a forecast. If the cumulative values for output shown by these two types of programme differ widely (see Fig. 3.5), it will be necessary to revise the annual programme and to modify any purchasing, financial, or other plans which are based on that programme.

3.5.4 *Shop-order progressing*

At the ordering level—the second level of production control—progressing again attempts to constrain events to follow the plans made when ordering with multicycle ordering. This is done by checking all orders at regular intervals by listing any incomplete orders, and by feeding back information about those for which the due date has already passed or is soon due, so that corrective action can be taken.

The order progress record normally takes the form of a list of overdue parts (see Fig. 3.6). One way of producing this list with multicycle ordering systems has already been illustrated in Fig. 1.11 where the list forms part of the dispatching and operation scheduling system.

With multicycle ordering systems, different part orders have different order quantities and order frequencies, and they also, therefore, have different order

Overdue orders (Made parts)			Group number	Week number	Overdue date	Due date
			2	22	21 May 1993	4 July 1993
Ordered			Produced		Action	
Part number	Name	Quantity	Done	Short		
A 2070	Filter bracket	30	27	3	Obtain missing material	

FIG. 3.6 A list of overdue parts for short-term programmes.

dates and due dates. These must be checked individually for each order against today's date to prepare the overdue list. With single-cycle ordering systems, such as Period Batch Control (PBC), sets of parts are ordered before the beginning of each period for completion by the end of the period. All parts on each overdue list have the same order and due dates. With PBC, the overdue-orders list is a crash signal. It is only needed in the event of a failure to complete all the parts in a list order by the due date.

When multicycle systems are used to regulate parts manufacture for assembly, it has been found from experience that shop-order progressing is not sufficient on its own. To ensure that complete sets of parts are available when needed for assembly, it is also generally necessary to have a supplementary *shortage-chasing system*. Stocks are checked against a requirement schedule some time before assembly is due to start, to find which parts are available, and a *shortage list* is prepared, listing any missing parts. A *progress chaser* or *expediter* then looks for the missing items, and outstanding operations are given priority to get them completed quickly.

With single-cycle systems, parts are made in one period for assembly in the next. Progressing involves checking that all the parts ordered in each period are completed by the end of the period. The need for reliability is much more stringent in this case, but there is no need for supplementary shortage chasing.

Call-off overdue list				Supplier: AB Foundry Ltd	Week number: 21	Call-off 21 May 1993	Due 28 May 1993
Part number	Name	Quantity call-off	Quantity received	Action			
A2070	Filter bracket	30	27	Visit supplier			

FIG. 3.7 Purchase call-off for late deliveries.

3.5.5 *Purchase-order progressing*

With purchase orders, the progress record again takes the form of periodic lists of overdue orders. Once again, the method of progressing depends on the type—multicycle or single cycle—of ordering.

With multicycle systems, different parts have different order quantities and frequencies, and different order dates and due dates. With single-cycle systems, different parts may again have different order quantities and frequencies, but they share a common series of call-off delivery dates. The purchasing method used in this case is known as the *call-off* method. Deliveries are called off from suppliers against purchase contracts at regular intervals, to meet production needs. In this case, progressing takes the form of a list at the end of each period, showing outstanding deliveries (see Fig. 3.7), all of which are followed up by a phone call or a visit.

3.6 Loading

The load and the capacity in factories are both measured in units of work, of man-hours and/or machine-hours. The load is generated in workshops by the issuing of programmes and orders. It starts as illustrated in Fig. 3.8 at the maximum value when the order is issued (OD), and it falls to zero when the order has been completed (DD). The feedback control used to constrain orders to follow the plans generated by programming and ordering is called progressing, this has already been described. The load cycle, in this case, is directly related to the order cycle.

Here, the terms *loading* or *load control* are used in the special sense of a control which measures and compares load and capacity, to ensure that there is sufficient capacity available to meet the requirement imposed by the load.

3.6.1 *Loading and assembly*

Consider, first, loading in assembly departments. Since assembly is mainly a manual process at present, the load and the capacity are generally measured in man-hours. If there is a high degree of flexibility in the workforce, so that most assemblers can do most assembly tasks, capacity is a simple function of the number of assemblers employed. The calculation of capacity is more difficult, however, if there is a high degree of specialization among the assemblers.

The assembly load per product or assembly can be measured in man-hours, and the total load imposed by a programme is then easy to calculate. The loading system must be planned to feed back information to the appropriate manager about cases where the load exceeds the available capacity. With the increasing mechanization and automation of assembly, the method used for assembly loading will tend to become the same as that for machining.

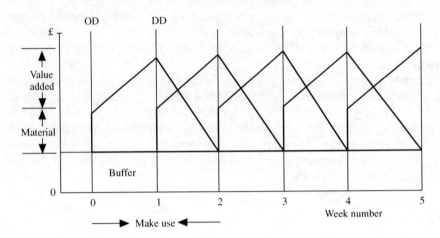

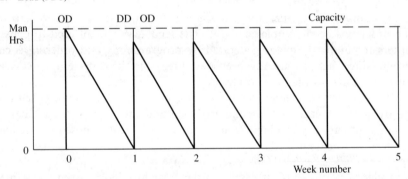

FIG. 3.8 Load variation with time: OD, order date; and DD, due date.

3.6.2 *Loading and machining*

The net capacity of a machine is the time for which it is available to do useful work. The gross capacity may be (say) 55 machine-hours per week (40 hours plus 15 hours overtime). This is reduced by lost time caused by machine breakdown, absentee workers, setting up, cleaning, and other causes, leaving a smaller *net capacity* available for productive work.

The *net load* on a machine is the sum of the operation times for all the jobs loaded on the machine. For some purposes, such as for some types of operation scheduling, it is necessary to know the *gross load* or the total time which a part spends on a machine, adding allowances for setting up, cleaning, machine breakdown, absentee workers, etc. to find the gross load.

THE PRODUCTION-CONTROL FEEDBACK CONTROLS 55

Two methods of loading are in use. First, the *gross-load method* compares the gross load with the gross capacity. Secondly, the *net-load method* compares the net load—or the sum of the operation times—with the net, or corrected, capacity. The net capacity in this second case, is found by random observation studies, or by machine-monitoring methods, which record when the machine is running and rely on supervisors or workers to record the reasons for any stoppages.

The three main problems with the gross-load method are: firstly, that setting-up times depend on the sequence of loading; secondly they vary enormously for the same operation on different occasions; thirdly, and it is much simpler to measure the net machine capacity than it is to adjust the net load, or a sum of the operation times, for all the many different parts which used machine, when trying to find their gross loads. For these reasons the net-capacity method is generally preferable. This view is reinforced by the fact that the latest methods of operation scheduling are less dependent on accurate gross-load-operation time data than the older methods.

It will be realized that loading figures have little meaning unless they are related to a specific period of time. With GT and a single-cycle system, such as PBC, the load imposed by a list order on a group arises during one period. With a multicycle ordering system, a similar approach, which attempts to measure the load imposed by all the outstanding orders, does not provide an accurate measure of future period loads. The total load at any moment is also much more difficult to measure, because different orders have different start and finish dates.

Figure 3.9 gives a simple example of PBC, with a three-period schedule of machining (M), of assembly (A), and of sales deliveries (S). If the operation time per part on each machine and the assembly time per product are known it is very simple to calculate the total-period net loads, and to compare them with the net capacities. It is generally safe to assume with GT and PBC that if the capacity is greater than the total load at a work centre it will generally be possible to find an operation schedule, which will complete the load by the end of the period.

3.7 Summary

The controls used in management to constrain events to follow plans are feedback controls based on systems theory. Control is obtained by monitoring selected system-output-variable values, by comparing these values with the plans made by management, and by feeding back information about major variances between them to managers so that they can take the action necessary to bring achievement back into line with the plans.

The three main feed-back controls in the function of PC are progressing, loading and inventory control. This chapter covered the first two controls, leaving inventory control for the next chapter.

Progressing constrains the product and component output to follow the plans contained in programmes and orders. Loading compares the load and

1. PBC schedule

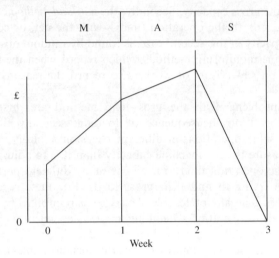

2. Load on a machine

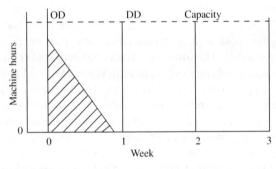

With PBC there must be sufficient capacity to cover the load. In other words the load must be zero by the end of the period.

3. Load assembly

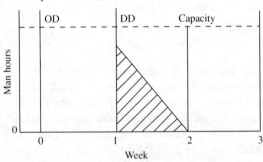

FIG. 3.9 Load, machining, and assembly for PBC: M = make parts, A = assemble; S = deliver to customer; OD = order date; and DD = due date.

capacity, and it feeds back information about any overload so that it can be corrected.

4
STOCKS

4.1 Introduction

One of the main objectives of Period Batch Control (PBC) is to regulate the flow of materials through factories, so that the investment in stocks is maintained at a low level. Reducing the stocks in a company first increases its flexibility, or its ability to follow changes in market demand without losses due to obsolescence, and secondly it reduces both the investment in stocks and also the cost of stock holding. Reducing the stock-holding costs increases the profit, so a reduction in the investment in stocks has a geared effect in increasing the rate of return on investment, or the *profitability*.

In order to understand the economic advantages of PBC, it is essential to understand the nature of stock, and the way in which its value can be raised or lowered in practice. This chapter starts by defining stock. Then the three main categories of stock, cycle stock, buffer stock, and excess stock, are defined. Next, the way in which changes in the throughput time, the run quantity, and the delivery batch quantity affect the cycle-stock level are examined, and then the factors affecting buffer stocks and excess stocks are discussed.

Up to this point, only a simple linear relationship between stocks and variables, such as output, throughput time, and run quantity, are considered. With some types of production there are also much more complex dynamic changes which distort these linear relationships and have a major effect on the stock level. These dynamic relationships were described in Chapter 1, and are considered again near the end of this chapter, which ends with a description of *inventory control*—the feedback control system used to constrain stocks to remain within preselected limits.

4.2 Stocks

Stocks are defined as: all the tangible (touchable) assets of a company, other than its fixed assets. Stocks do not include: land, buildings, processing machinery, service plant, or special tooling. *Stocks* do include all bought and made materials, finished parts, finished assemblies, and finished products, held in and out of stores. They include work-in-progress (WIP), which consists of materials at various stages of conversion between the category change points above. Stocks also include consumable tools and general materials, such as paint, lubricating oil, greases, and welding rods.

INTRODUCTION

Manufacturing companies invest some of their capital in fixed assets. The rest is *circulating capital*, which is used to pay wages, to pay for materials and other purchases, and for other forms of recurring expenditure. The value of the circulating capital is reduced by these payments for wages, materials, and other expenses, and is increased by the addition of revenue received from the sale of goods. An excess of these revenues over the expenditure is called the profit. At any moment in time, most of a company's circulating capital is tied up (invested) in stocks.

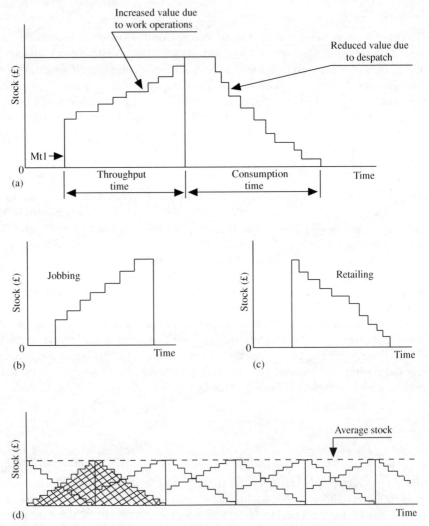

FIG. 4.1 Stock chart model of changes in stock value with time: (a) a stock chart for a manufactured item; other typical stock charts include (b) jobbing, (c) retailing, and (d) stock charts in series to represent continuous supply.

A very large part of the capital invested in British manufacturing (fifty per cent) is invested in stocks. A much smaller proportion (estimated as thirty five per cent), is invested in stocks by Japanese manufacturing industries. This difference represents a major factor in explaining the present economic superiority of the Japanese in manufacturing.

The changes in stock with time can be illustrated by stock charts (see Fig. 4.1). The changes in stock value during manufacture and consumption usually take place as a series of steps, but can be conveniently illustrated (as in the remaining figures in this chapter) by straight lines instead of steps.

Stocks can be classified into three main types: cycle stock, buffer stock, and excess stock (see Fig. 4.2).

1. *Cycle stock* is the fluctuating value of the stock induced by manufacture or purchase in batches. It rises with the manufacture of batches of made parts or with the delivery of batches of bought parts, and it then falls with the consumption of these batches. Continuous line flow (CLF) can be seen as the limiting case where the *transfer* quantity of materials between operations is equal to one.

2. *Buffer stock* is an arbitary addition to stocks, which is provided to insure against late delivery or scrap (insurance buffer), or to compensate for seasonal or random variations in demand (smoothing buffer), or to reduce delivery times (delivery buffer).

3. *Excess stock* is unnecessary stock. It is usually generated by accident. It is a form or waste, because most of it will end as obsolete material, if and when product ranges or designs are changed.

One further category of stock—which is not in the same general classification as cycle, buffer, and excess stock—is *obsolete stock*. Obsolete stock is stock which, due to such reasons as a change in the product range or in the product

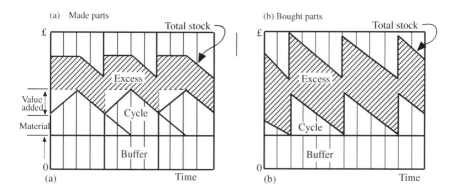

FIG. 4.2 Types of stock (cycle, buffer and excess stock) For: (a) made parts, and (b) bought parts.

design, has lost all its value except its scrap value. Most excess stocks eventually become obsolete stocks.

4.3 Cycle stocks and related variables

We will look first at the cycle stocks of made parts. The four main variables whose values determine this cycle-stock value are product output, run quantity, run frequency, and throughput time, where:

1. The *output*, O, is the number of products produced per year, or in some other specified period of time.
2. The *run quantity*, RQ, is the quantity of a product or part processed as a batch at a work centre before changing to make some other product or part, or before stopping the machine and leaving it standing idle.
3. The *run frequency*, F, is the number of runs of a product or part, produced per unit of time (for example, per year).
4. The *throughput time*, T, is the elapsed time taken for the conversion of materials into parts or products for a particular product, part, or run quantity, and for a specified segment of the material flow system (MFS).

It can be seen that:

$$O = RQ \times F \qquad (1)$$

$$F = \frac{Q}{RQ} \qquad (2)$$

and that T is a function of RQ.

The run quantity RQ, is a type of batch quantity. There are many ways in which parts or products can be batched for convenience in production. Three other ways which are of special importance in production control (PC) are:

1. The *order quantity*, OQ. This is the number of products or parts authorized for manufacture or purchase by the issue of an order.

2. The *set-up quantity*, SQ. This is the number of parts—not necessarily all the same—which are made on a machine between changes in the tooling set-up. (Different, but similar, parts can often be made one after the other at the same set-up).

3. The *transfer quantity*, TQ. This is the number of parts transferred as a batch between the machine tools (or other work centres) used for successive operations. The limiting values for TQ are: one (minimum) and the value of the run quantity, RQ (maximum).

These three variables, together with the run quantity, are different types of *batch quantity*. All four are independent variables. Within certain limits, the value of any one of these variables can be changed without changing the value of the other three (see Fig. 1.9).

Changes in the value of these four types of batch quantity have different economic effects. It is important to note that if the value of one of these four variables is changed—leaving the other three variables unchanged—then only the run quantity directly affects the value of the cycle stock.

4.4 Reducing cycle stocks

It was stated in the previous section that changes in the output, the run quantity, the run frequency, and the throughput time all affect the value of the cycle stocks for made parts (see Fig. 1.10).

4.4.1 *The run quantity and frequency and the cycle stock for made parts*

The relationship between the run quantity and frequency and the cycle-stock value is illustrated in Fig. 4.3. At a constant output, reducing the run quantity will increase the run frequency, and it will also reduce the average value of the cycle stock at the same time.

Under the same condition of a fixed output rate, increasing the run frequency will reduce the run quantity, with the same effect of a reduction in the value of the cycle stock. If the output changes this must change either the run quantity and/or the run frequency, and depending on the direction of change induced in these variables the value of the cycle stocks will again change.

4.4.2 *The throughput time and the cycle stock for made parts*

The relationship between the the throughput time for made parts and the cycle-stock value is illustrated in Fig. 4.4. At a constant output, and with fixed values for the run quantity and frequency, reducing the throughput time will reduce the value of the cycle stock because it reduces the number of different batches in production at the same time.

The throughput time is mainly a function of the complexity of the MFS (see Fig. 2.1). It is also, however, partly a function of the run quantity. Reducing run quantities reduces stocks in their own right (as shown in Fig. 4.3), but it also achieves a further reduction by reducing the throughput time (as illustrated in Fig. 4.4).

The run quantity is a parameter, that is a variable to which a manager can assign an arbitary value at will. The throughput time, on the other hand, is a *system variable*. Its value can only be changed indirectly, by changing the value of related parameters. In practice, the most effective ways of reducing throughput times are:

1. *Group Technology* (GT). This brings all the operations to make a part together in the same group, and greatly simplifies the MFS (see Fig. 2.1).

2. *Close-scheduling*. Where the following operations on a batch of parts are started before the preceding operations have been completed for all the parts in the batch (see Fig. 4.5).

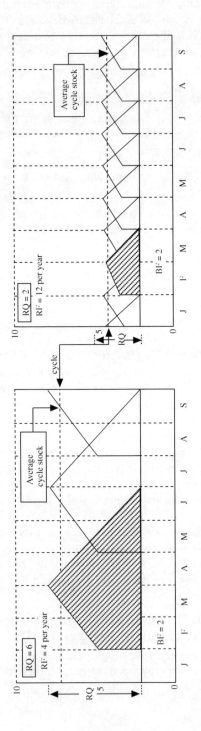

Fig. 4.3 Cycle stocks and the run quantity (made): BF = buffer stock, RQ = run quantity, and RF = run frequency.

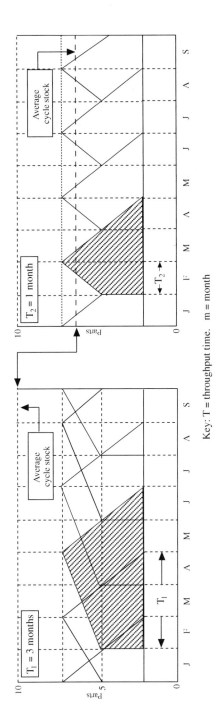

Fig. 4.4 Cycle stocks and the throughput time, (made).

(a) Open scheduling

Machine		Day number																										
Letter	Name	1	2	3	4	5	6	7	8	9	10	11	12	13	14	15	16	17	18	19	20	21	22	23	24	25	26	27	28
A	Lathe	Op 1																											
B	Lathe				Op 2																								
C	Universal mill								Op 3																				
D	Drill													Op 4															
E	Tap																						Op 5						

Throughput time = 23 days

(b) Close scheduling

Machine		Day number																										
Letter	Name	1	2	3	4	5	6	7	8	9	10	11	12	13	14	15	16	17	18	19	20	21	22	23	24	25	26	27	28
A	Lathe	Op 1																											
B	Lathe		Op 2																										
C	Universal mill			Op 3																									
D	Drill				Op 4																								
E	Tap					Op 5																							

Throughput time = 4½ days

FIG. 4.5 Open and close-scheduling and the throughput time.

3. *Reducing the run quantities.* With PBC the run quantity is a function of the ordering period. To reduce the run quantity the common ordering period must be reduced.

4. *Reducing the operation-scheduling times.* The operation times, plus set-up times, plus the down time, plus other idle times can be reduced.

5. *Adding shift working.* This increases the capacity without increasing the elapsed calendar days, or, in other words, the throughput time.

4.4.3 *Reducing the cycle stock of purchases*

The cycle stock of purchased items is a function of the delivery batch quantity and the delivery batch frequency, as illustrated in Fig. 4.6. To reduce these stocks at a constant output rate, the delivery batch quantities must be reduced and the delivery frequency must be increased.

In practice, this reduction will normally require a change from multicycle ordering and batch buying, to single-cycle ordering and the call-off method of purchasing. In this latter case, all the items are delivered as required, to meet call-off instructions issued at regular period intervals. All call-off instructions issued on the same date will have the same due date.

It can be seen that the relationship between the cycle stock and its related variables is approximately linear and that it is relatively simple. If the need for buffer stock could be eliminated or reduced, if the generation of excess stock could be eliminated and if the dynamic generation of stocks could be eliminated (see Section 4.7), then stock regulation should become a relatively simple task, which is only concerned with cycle stock. In practice, we can never hope to achieve this ideal.

4.5 Buffer stocks

The buffer stock is a parameter, or a variable to which a manager can ascribe arbitary values at will. To reduce buffer stocks therefore, it is only necessary that the manager should reduce them. This solution, however, ignores the fact that buffer stocks were introduced to solve problems. It might be said, therefore, that the best way to reduce buffer stocks is to eliminate the problems which they were introduced to solve. We will examine, in turn, the problems involved in regulating insurance, smoothing, and delivery buffer stocks.

4.5.1 *Insurance-buffer stocks*

Insurance stocks are provided to insure against accidents, such as late delivery by a supplier or scrapped work during processing. The best way to eliminate the need for insurance stocks is to improve the reliability of manufacturing units and of suppliers.

The general approach to reliability improvement can be seen as a process of control, or of comparing the actual output with the planned output, followed by the feedback, to the management of information about the variances

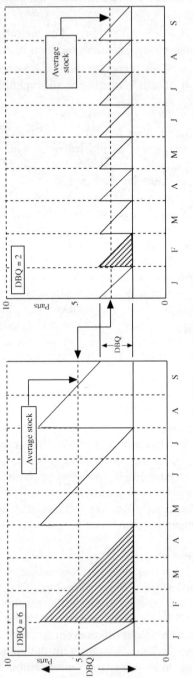

FIG. 4.6 Cycle stocks and the purchase delivery batch quantity (DBQ).

between them, followed by correction. In the case of manufacturing, GT has much to offer in the improvement of reliability. With traditional process organization, the process-specializing sections do not complete parts. They cannot, therefore, be held responsible for the quality of parts or for the completion of parts by their due date. With GT, the groups complete all the parts they make, and full accountability is possible at shop floor level. Some idea of the potential for the improvement in the manufacturing reliability in the West can be seen in the provision in some Western companies of a scrap allowance of the order of ten per cent (100 000 parts per million), compare this with the values in Japan, where in the better companies the scrap is measured in rejects per million parts produced, normally in quantities of less than ten rejects per million.

In the case of purchasing, the need is for closer contact with suppliers. For economic reasons, this tends to lead to single sourcing to reduce the number of suppliers, and to single-cycle ordering to reduce the number of different order dates and due dates.

For most of the made parts, and for at least some purchased parts, spare capacity can be provided as a substitute for insurance stocks. With this approach, missing or defective parts are replaced either by new manufacture or by special deliveries from suppliers and not from buffer stock. Considering that in most companies only the bottle-neck machine can be fully loaded efficiently, it will be realized that much of the capacity needed for this replacement approach is already available. If a reasonably high level of manufacturing reliability has been reached, a special night shift can probably overcome the occasional disaster more efficiently and economically than is possible by holding insurance stocks for all the parts being manufactured. The elimination of all insurance buffer stock is always a worthwhile objective.

4.5.2 *Smoothing buffer stocks*

Smoothing stocks are stocks of finished products, generated by methods used to conserve capacity, in industries which make standard products and which are subject to random or seasonal variations in demand. Figure 4.7 shows the general approach in the case of random variations. The capacity level of the factory is maintained at the same level as the mean rate of demand. Using the PBC method, if the sales orders in a period are below the capacity level additional products are ordered for stock (to use all the available capacity). In later periods, when the demand is greater than the capacity, the output provided by the available capacity is supplemented by drawing on the smoothing stock to meet the demand. A maximum permissible level of smoothing stock is normally fixed, to avoid the risk that a sudden decrease in orders may lead to an excessive rise in stocks.

In the case of *seasonal* variations in the product demand, smoothing stocks have to be planned as part of the planning of the annual programme. Statistical analysis of past sales can be used to determine the percentage of the annual

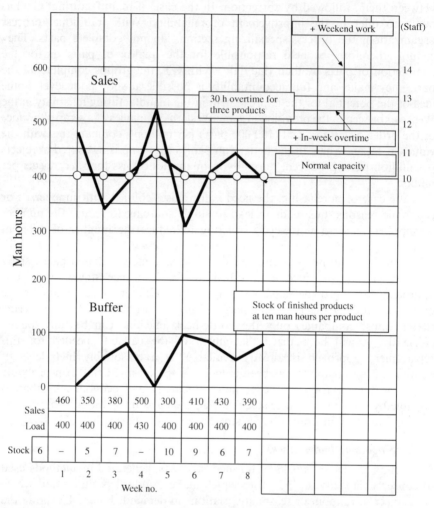

Fig. 4.7 Assembly capacity and smoothing.

sales which is normally sold each period (see Fig. 4.8). Sales, production, and stock programmes are then planned to give an approximately even rate of production. In some factories, seasonal variations in the capacity (overtime or contract labour) are used to compensate for major variations in the seasonal demand. Factories may make some products for stock in the low-demand season, and they may also use the lower level of capacity. They can then increase the capacity to a maximum during the high-demand season, and at the same time they can use any accumulated stocks to help meet the market demand.

If the products produced have a very seasonal sales curve, it may be desirable to modify the range of products to produce a *seasonally balanced range*, which

Sales	Product: X				Forecast: 2600 per year				Year: 1993			
Month	January				February				March			
Sales per week	40	40	40	40	45	45	50	50	50	50	50	50	50
Week number	1	2	3	4	5	6	7	8	9	10	11	12	13
Percentage of annual sales	1.54	1.54	1.54	1.54	1.73	1.73	1.92	1.92	1.92	1.92	1.92	1.92	1.92
Month	April				May				June			
Sales per week	50	50	50	55	60	60	60	60	–	60	60	60	70
Week number	14	15	16	17	18	19	20	21	22	23	24	25	26
Percentage of annual sales	1.92	1.92	1.92	2.12	2.31	3.31	2.31	2.31	–	2.31	2.31	2.31	2.69
Month	July				August				September			
Sales per week	70	70	70	70	65	60	55	55	55	50	50	50	50
Week number	27	28	29	30	31	32	33	34	35	36	37	38	39
Percentage of annual sales	2.69	2.69	2.69	2.69	2.5	2.31	2.12	2.12	2.12	1.92	1.92	1.92	1.92
Month	October				November				December			
Sales per week	50	45	45	45	45	45	45	45	45	45	40	40	–
Week number	40	41	42	43	44	45	46	47	48	49	50	51	52
Percentage of annual sales	1.93	1.73	1.73	1.73	1.73	1.73	1.73	1.73	1.73	1.73	1.54	1.54	–

FIG. 4.8 The seasonal distribution of the annual sales.

can be manufactured continuously throughout the year with little stock (see Fig. 6.8).

Random and seasonal variations in the sales demand are very difficult to eliminate, and if capacity is a problem smoothing is often the most economical way to conserve the capacity. Care is needed, however, to ensure that the amount of smoothing buffer stock is monitored and controlled.

4.5.3 *Delivery buffer stock for assembled products*

Delivery buffer stocks are stocks of products which are held to improve the delivery performance to customers. In some markets products can only be sold if they can be delivered immediately on demand from stock. In other markets this service is needed for some sales orders but customers for other products are prepared to accept longer delivery periods.

If delivery buffer stocks are used, products are taken from the delivery stock and are sent to customers immediately on receipt of their orders. A standard level of delivery buffer stock is fixed for each product. All the orders received during each period are listed, and then ordered in these exact quantities for manufacture in the period which immediately follows. They are sent on completion to the delivery buffer store, replacing products sent to customers from stock in the previous period.

In this method, as illustrated in Fig. 4.9, orders are made in the exact quantity needed in each period to replace all the issues from store, that is, to replace all the orders received and delivered in the previous period. The standard level of delivery buffer stock in the stores is only reached if there is no demand for the product for a number of periods. If all products are in regular demand, the delivery-buffer-stock level in the factory will generally be considerably smaller than the sum of the standard delivery-buffer-stock levels for all products.

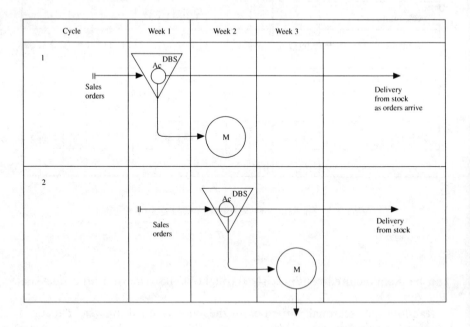

FIG. 4.9 Delivery buffer stock: AC = accumulate a list of the orders, DBS = delivery buffer stock, M = make parts.

4.5.4 *Delivery buffer stock for parts*

In implosive and square industries, where products take the form of single-piece parts, delivery buffer stocks of parts may be held in the same way as that described for assembled products.

There are also some circumstances in which it may pay to hold delivery buffer stocks of parts for assembled products. If, for example, the product range consists of a standard product, sold in say twenty different variants, each formed by adding a small number of special parts to the standard list, it may pay to hold buffer stocks of these special variant parts. If this is done, standard products can be planned for continuous production. The allocation of product variants to customers, and their assembly, then only needs to be planned at the beginning of the assembly week, giving say, a delivery time of, say, two weeks to customers instead of four weeks. The investment required to give a two-week delivery in this case, by holding sets of parts for variants, would be much less than would be the case if the buffer stock were held in the form of finished products.

Just-in-time (JIT) production with the elimination of all delivery buffer stock is again a worthwhile objective in the majority of cases.

4.6 Excess stock

Excess stock is stock in excess of the quantity needed to sustain production, that is, it is the total stock less the sum of the cycle stock induced by the actual sales plus the buffer stocks. It is not difficult to measure the excess stock for a single product type. Smoothing buffer stocks and delivery buffer stocks are normally held in the finished-product form, and, if there is no insurance buffer, the cycle and the buffer stocks are easy to count. The total excess stock is the sum of the excess stocks for all the different products in production. This total is difficult to measure with a multicycle system of ordering, but it is relatively simple to measure with a single-cycle system.

Excess stock arises due to ordering errors. If more products are ordered than are sold, excess stocks will be generated. Figure 4.10 shows a case where orders based on explosion from a long-term programme Materials-Requirement Planning (MRP) produced twice as many products as were sold. Each successive order induced a major increase in the excess stock.

In the case of component processing, the widely held belief that machine utilisation is a reliable indicator of manufacturing efficiency is a common cause of excess stock. In fact, however, only the bottleneck machine in a factory can be fully loaded economically. This machine dictates the level of the product output which can be achieved. If machines, other than the bottleneck, are loaded beyond this output level, they produce excess stock, and there is a high probability that this will eventually become obsolete.

Period batch control assembles, in each period, the number of products needed for distribution to customers in the next period, plus any minor

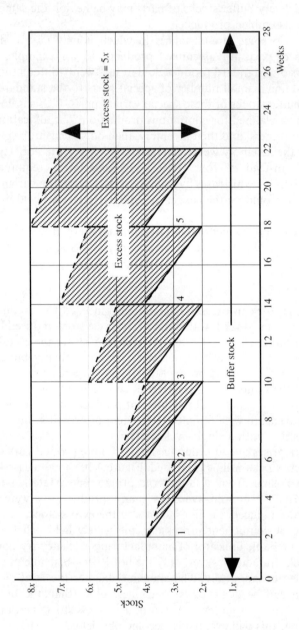

FIG. 4.10 Excess-stock generation. Note, the orders per two-week period = $2x$, and the consumption = x.

addition for adjustment of the buffer stock. Normally, in PBC, in each period, only the number of parts needed for assembly or for distribution in the following period are made. Therefore, PBC is an ordering system which does not produce excess stock.

4.7 Dynamic changes in stocks

The changes in stock considered up to this point have been governed by relatively simple linear relationships. One might well ask: 'If PC is as simple as this, why has industry been so unsuccessful in designing reliable systems to regulate and control materials flow in factories?' The answer seems to lie in the dynamic effects which distort the simple linear stock relationships described above. We must now look at these dynamic effects and see if it is possible to avoid them. These dynamic changes are induced mainly by the PC methods chosen to regulate materials flow through the MFS. Three of these dynamic changes will be examined here:

(1) interference;

(2) the surge effect;

(3) the demand-magnification effect;

4.7.1 *Interference*

The straight-line relationships illustrated in the stock charts shown in Fig 4.3 and 4.4 are very seldom consistent in practice if process organization and multicycle ordering are used. Differences in the product mix and in the loading sequence, leading to differences in the queueing time and in the load and capacity, can cause the throughput time to be different every time a part is made.

One can hope to achieve some consistency in operation times, particularly with NC machining, but the total throughput time, including the setting-up and the queueing time, will tend to vary. Setting-up times, for example, vary with the sequence in which parts are made on a machine. Changes in the scheduling sequence will affect the value for the total setting time; and they will, therefore, also affect the load on the different machines used. These changes in the capacity and the load will affect the queueing times and they will therefore also affect the throughput times.

A big improvement in the throughput-time consistency is possible with CLF. An important improvement is also possible with GT coupled with PBC. Consistent throughput-time control is impossible with process organization and multicycle ordering.

4.7.2 *The surge effect*

The second of these dynamic effects, the surge effect, was described in Chapter 1, and it was illustrated in Fig. 1.7. If different parts are ordered with different order frequencies, the intervals of time between the peaks and troughs of their stock and load cycles will be different. As time passes, the peaks and troughs of the many different cycles, for different parts, will drift into and out of phase, causing large and completely unpredictable *surges* (rises and falls) in the stock level and in the balance of the outstanding load.

This surge effect only occurs with multicycle ordering systems. (See Fig. 1.7). It can be completely eliminated by changing to a single-cycle system such as PBC.

A close analogy to the surge effect can be found in electronics. If signals with different wavelengths are transmitted through a communication channel connected to a loud speaker, they generate 'noise' at random intervals when the peaks and troughs of the different signals drift into phase. The surge effect provides one of the main reasons for the poor performance of most present-day PC systems.

4.7.3 *The demand-magnification effect*

The third of these dynamic effects is the demand-magnification effect, it is sometimes called the industrial-dynamics effect or the J. W. Forrester effect. It has been observed in industry over many years that when demand is passed through a series of independent inventories (as, for example, in the case of a distribution channel connecting customers, retailers, wholesalers, and a factory) the demand variation gets bigger with each successive transfer (see Fig. 1.6).

This magnification effect was simulated by J.W. Forrester at the Massachusetts Institute of Technology (MIT) in 1962, in his research entitled *Industrial dynamics*. Forrester showed that the variation in demand induced a similar variation in the stock levels in each inventory. He believed that these variations provided a major reason for the economic cycles between booms and slumps in national economies.

Forrester believed that one reason for the magnification effect was the wide use of stock control ordering systems based on reordering levels. He found that a change to base-stock control—which is a stock-replacement system—would reduce the magnification. Period Batch Control (PBC) has the same effect, because parts are ordered in sets.

Research into by Professor Dennis B. Towill and his research team at the University of Wales (UWIST), is studying the magnification effect in large industrial companies (the logistical dynamics of supply chain management). Their early observations indicate that to eliminate the effect completely the total supply chain needs to be studied from materials supply through to the final customer. The problem is not limited just to the distribution chain, and magnification in any part of the material flow system can affect the whole.

4.8 Inventory control

Feedback control used to constrain stock levels to follow those planned by PC is known as *inventory control*. It should be noted that the term *stock control* is not used in this sense because this term is normally reserved as the name of a class of stock-based ordering systems in which ordering is based on stock reorder levels (see Fig. 1.5).

4.8.1 Inventory control for assembled products

To maintain an accurate record of the stock of assembled products, it is essential that there should be fixed locations for storage. It may be convenient, however, to store different classes of the same product in different locations. *Smoothing stock*, for example, may be stored in the assembly group which makes the product, and *delivery buffer stock* may be stored in the packing and despatch department.

If there is very little finished-product stock to be counted, the most accurate and cheapest method is to make a physical check on, say, Friday each week before the programme meeting. For accuracy, it is desirable that the check sheet should, in this case, list the serial numbers of all the products included in the record.

If there are large numbers of products to be counted, as in the case of highly seasonal products or for products for which delivery buffer stock is needed, then a running record is preferable, showing all receipts, issues, and the balance of stock after each transaction. The prime data used to make the record should be carefully specified. For example:

1. Products shown as being completed in the *daily final-inspection sheets* will be added to the record.
2. Products shown in *despatch notes*, as having been delivered to customers will be substracted from the record.

Records of this type should be subjected to regular physical checking. Feedback control is obtained by comparing the stock record with the planned stock levels shown in the stock programme (made during programming).

4.8.2 Inventory control for piece parts

It might be said that it is easiest to count stock when there is none of it. With multicycle ordering systems, it is difficult to avoid the need for controlled stores between processing stages. For example, in engineering-assembly factories, there will generally be:

(1) a casting store, between the foundry and machining;

(2) a forging store, between the forge and machining;

(3) a finished-parts store, between machining and assembly; and so on.

In all these cases, it will be necessary to maintain running stock records for all items. To use such records as the basis for a feedback control of stock levels maximum stock levels must be fixed for all items, and there must be feed back when these levels are exceeded. This seldom provides a satisfactory control, because demand is constantly changing, and it is impractical to change the maximum stock values as frequently as demanded by the market requirements.

Major reductions in stock levels are possible with GT and PBC. First, with GT it is often possible to combine following process stages in the same group. Examples from practice include:

(1) groups which both make die-castings and machine them, thus eliminating the need for a casting store;
(2) groups which both make parts and assemble simple products (examples include welded fabrications and the manufacture of hand tools)—this combination eliminates the need for a finished-parts store.

Secondly with PBC, castings made in one period will be machined in the next period, and they will be assembled in the following period. At the end of each period, the stock remaining from the production in that periods should be zero, and therefore it should be easy to count. Once again, there is no need for a running record. A physical count of uncompleted jobs, at the end of each period, will give both progress control and inventory control.

Finally, in the case of purchased items, if the call-off method of purchasing is used, most materials and parts will be scheduled to arrive in one period, ready for further processing, or assembly in the next period. A progress control based on call-off instructions can again provide, at the same time, a control of the inventory and of purchasing.

The main exceptions, where a running record will still be needed for piece parts, will cover such items as:

(1) spare parts for sale;
(2) buffer stocks;
(3) purchased items, where it has not yet been possible to use call-off.

4.8.3 *Reconciliation with accounting stock*

A principle objective of inventory control is the provision of a means for checking capital tie-up and credit performance. One needs:

(1) a list of the stocks in the factory at a given moment in time, and a figure for the total value for all the parts;
(2) a figure for the accounting stock value, which is found by adding the revenues received for sales and subtracting the expenditure incurred in manufacture.

To reconcile these two figures, the main problem is the treatment of the credit taken on purchases and that allowed on sales. One company achieves this reconciliation, in a simple case, by maintaining separate records for the following classes of accounting stocks:

(1) purchases received and not yet paid for (A);
(2) purchases received and paid for (B);

(3) sales despatched and not yet paid (C);

(4) sales despatched and paid for (D).

If both purchases and sales are valued in the same units of cost, the physical stock record should equal $(A + B) - (C + D)$. The accounting stock record equals $(B - D)$.

4.9 Summary

Stock is all the tangible assets of a company other than its fixed assets. There are three main types of stock: cycle, buffer, and excess stock.

The level of cycle stocks is determined partly by changes in the value of a relatively small number of system variables—notably the output, the run quantity and frequency, and the throughput time. The level is, however, also affected by dynamic influences which are much more difficult to evaluate, but which can be eliminated by choosing suitable production systems.

The regulation of stock has been a major problem in manufacturing since the start of the Industrial Revolution. The main difficulties have been caused by a tendency to see stock regulation as a very simple problem. It has only recently been understood that stock regulation is in fact a very complex problem.

The choice of PC system has a major influence on the stock investment. Period Batch Control is particularly favoured in this respect because:

1. It works with very little stock, and the product range can be changed with little obsolescence.
2. It does not produce excess stock.
3. It tends to eliminate the surge effect.
4. It greatly reduces the magnification effect.
5. It is simple to operate and control.

PART 2

PROGRAMMING, ORDERING, AND DISPATCHING WITH PBC

PART 3

5

THE DATABASE FOR PBC

5.1 Introduction

The database needed for production control (PC) is studied in this chapter, with particular reference to the data needed for Period Batch Control (PBC). Production-control data can be classified into seven main types:

(1) plant lists;
(2) lists of workers;
(3) process routes (route cards);
(4) parts lists (bills of materials);
(5) operation sheets;
(6) lists of suppliers;
(7) lists of customers.

The first three categories are always needed. The fourth category (parts lists) is only needed for complex assembled products. With simple assembled products, the necessary parts-list data can often be included in the process routes. Operation sheets, which give detailed information about particular operations listed in process routes, are normally only used for the more complex types of part. Finally, there are some types of production control in which lists of customers and/or suppliers are needed. This chapter considers each of the above categories in turn, and then looks at the problem of how to maintain the accuracy of the database.

5.2 The plant list

A *plant list* is a list of all the machine tools and the other manufacturing facilities available in a company. A typical plant-list pro forma for a company making mechanical-engineering products is illustrated in Fig. 5.1.

5.2.1 *Plant-type code, plant number, and name*

The first three columns in Fig. 5.1, give a type-code number, a plant number, and a name for each type of machine tool, and for the other production facilities available in the factory. Ideally, the type code would be so designed that any operation allocated to one machine with a particular code number could be

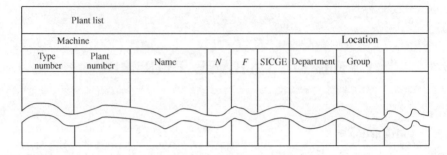

FIG. 5.1 A plant list. N = no. of machines of each type; F = no of different parts with operations on a machine.

transferred without difficulty to any other machine with the same number. In practice, there are so many transfer reservations due to machine modifications, size, condition, and ancillary equipment that it is safer to look at the type codes as short lists of machines, between which operation transfer may be possible.

Operations can be transferred between machines of the same general type, even if they differ in their specific type. For example, turning operations can be transferred between centre lathes, turret lathes, and turning centres. It helps if the code first indicates the general type (for example, T for turning), and then the specific type (for example, C for centre lathe).

A typical plant-list type code is illustrated in Fig. 5.2. The first character in this code indicates the general type of plant. The second character divides each category in this primary class into its own special subcategories (turning machines and drilling machines are used as examples), and the third and fourth characters specify each machine of each type with numbers, that is, 01–99. This code not only indicates the type of machine, but it also provides a specific *plant number* which can be used to identify each machine in the factory. For example, TC05 is a particular centre lathe installed in the factory.

In most existing factories, the plant number is not derived from a type code. In many cases, it will have been taken from a numbered plant register, arranged in the sequence in which machines were purchased. These numbers will be included in all the existing process routes, and they will be widely remembered in the factory. Changing the numbering system can be a very confusing and expensive operation. If it is decided not to eliminate the old numbering system, the existing plant numbers must also be included in the plant list, as in Fig. 5.1. In this case, a type-code number was also used. For many of these type codes, there were two or more plant numbers.

5.2.2 *Information about each machine*

The information about each machine which is needed for PC includes its name values for N (big N), for F (big F), the SICGE code; and the location.

THE DATABASE FOR PBC

Plant type code					
First character		Second character turning machines (T)		Second character drilling machines (D)		Third and fourth characters
A		Ta	Copy lathe	Da		
B	Broach	Tb	Bar lathe	Db	Bench drill	
C		Tc	Centre lathe	Dc		
D	Drilling machine	Td	Capstan lathe	Dd		
E		Te		De		
F	Flame cutting	Tf	Face and centre	Df		
G	Gear cutting	Tg		Dg		
H	Heat treatment	Th	Horizontal borer	Dh		
I		Ti		Di		
J		Tj		Dj	Jig borer	
K		Tk		Dk		
L	Laser	Tl		Dl		
M	Milling machine	Tm		Dm	Multispindle	
N		Tn		Dn		
O		To		Do		
P	Planer	Tp		Dp	Pillar drill (one speed)	
Q		Tq		Dq		
R		Tr	Revolver	Dr	Radial drill	
S	Shaper	Ts		Ds		
T	Turning machine	Tt	Turret	Dt	Tapping machine	
U		Tu	Turning centre	Du		
V		Tv	Vertical borer	Dv		
W	Wire cutting	Tw		Dw		
X		Tx		Dx	Pillar drill (two speed)	
Y	Painting	Ty		Dy	Pillar drill (three speed)	
Z	Machine centre	Tz	Crankshaft L	Dz	Pillar drill (four speed)	

FIG. 5.2 A plant-type code.

> **S – Special category**
> There is only one of each S-type machine and it would be very difficult to transfer the work that it does to any other machine type in the factory, for example, a gear hobber or a crankshaft lathe.
>
> **I – Intermediate category**
> Similar to the S-type but there is more than one of the type.
>
> **C – Common category**
> There are several of each type of machine and it is easy to transfer the work they do to other related machine types, for example, most lathes, mills, and drills.
>
> **G – General category**
> There are few machines of the type. They are used for a high proportion of the made parts or for many different types of part. It is unlikely to be possible to include them in groups, for example, large automatic painting or electroplating plants.
>
> **E – Equipment category**
> These items are used to assist manual operations, for example, benches, surface plates, and hand power tools.

FIG. 5.3 The SICGE code.

1. N is the number of machines of a particular type. It is found by counting machines with the same code number (First and second characters).

2. F is the number of parts with one or more operations on a machine type. It is found by counting the number of parts with operations on each machine type in the process routes.

3. The SICGE code (see Section 2.6) divides the plant into five categories: S = special, I = intermediate, C = common, G = general, and E = equipment. These categories are based on the possibility for transfer of the operations done on a machine to some other machine. They are defined in Fig. 5.3. Their main use is for production flow analysis (PFA), which is the technique used to plan the division into groups and families for Group Technology (GT).

4. The location of each machine, showing where it is installed, is also needed.

The location will change with the change from process organization to GT. Figure 5.1 shows the situation after introducing GT, when each plant number has been allocated to a department and to a group.

5.3 The list of employees

A list of employees is illustrated in Fig. 5.4. This gives the number and name of each employed person, and it shows the types of machine which he or she is qualified to set up and to operate. As far as PC is concerned, this information is needed for loading—checking the load against the capacity measured in man-hours and for operation scheduling.

THE DATABASE FOR PBC

Personnel record							Group number: Name: Cl, Turn		
Employee			Machine						
Number	Name		Tt	Tc	Tu	Mh	Mu	Dp	Dt
1073	J. Andrews	O	√	√	√			√	
		S	√	√				√	
1081	D. Barnes	O	√	√				√	√
		S	√	√				√	
1101	P. Bentley	O	√	√	√			√	√
		S		√	√				√
1230	R. Douglas	O				√	√	√	√
		S				√	√		
1246	K. Foulkes	O		√		√	√	√	
		S				√	√	√	√
1374	J. Johnson	O	√		√			√	√
		S	√		√				√
1375	F. Morris	O			√	√	√	√	
		S						√	
1399	T. Tomkins	O		√				√	√
		S						√	√
		O							
		S							
		O							
		S							
		O							
		S							

FIG. 5.4 A list of employees. O = operate, S = set-up.

The record of employees in Fig. 5.4 is based on GT groups. Increasing the number of machines which each worker can operate and set up increases the *flexibility* of a group, making it possible to follow changes in product mix and in the market demand for different products without losing capacity or sales. It must be remembered that a worker may be very flexible in one group, but

86 PROGRAMMING, ORDERING, AND DISPATCHING WITH PBC

may have no flexibility if moved to another group with different machines. The list of employees also provides the data needed for planning future training.

A subsidiary purpose of the list of employees is to guard against fraud. It is not unknown for wages to be paid in large companies to non-existent workers. Accurate and frequently checked lists of authorized employees make this type of fraud more difficult.

5.4 Process routes

A process route is a list of the operations used to manufacture a part, on assembly, or complete product. Essentially, it provides a list of operation numbers in operation sequence, a name for each operation, and it quotes the plant number and the name of the machine or other facility used for each operation.

5.4.1 *Typical process routes*

Figure 5.5 illustrates a typical process route for a made part used in a mechanical engineering product, and Fig. 5.6 shows a route for an assembly used in the same product.

Process route			Part number: Part name:				
Materials							
Operation		Machine			Op. time for each part (mins)		
Number	Description	Type	Number	Name			
01							
02							
03							

FIG. 5.5 A process route For machining.

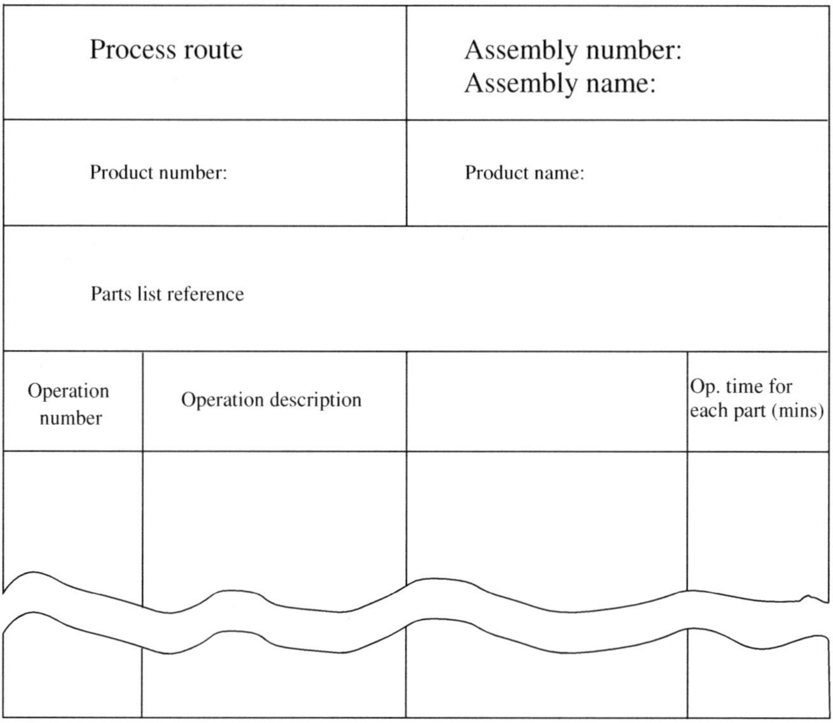

Process route	Assembly number: Assembly name:
Product number:	Product name:
Parts list reference	

Operation number	Operation description		Op. time for each part (mins)

FIG. 5.6 A process route for assembly.

Ideally, there should be a separate process route for each processing stage. In most companies making assembled products, this means one route for each part and one route for each assembly. Such routes should start by identifying the material to be used and end by indicating the department where the following stage will take place. If, for example, a route covers machining in a machining department, ready for assembly at the following stage, its process route should start with;

issue the following material from the material stores,

and it should end with a final operation,

transfer to the assembly department, say group A2, for inclusion in product type X.

As a second example, a route covering assembly and machining in a welded assembly department, might start,

collect together parts for assembly number XXX,

it may continue with a number of assembly operations, for example welding, riveting, brazing, and/or manual assembly, followed by machining, testing, and painting, and it may end with a final operation,

pack and dispatch.

5.4.2 Combined process routes and parts lists

With complex products, it is generally necessary to divide component processing and assembly into separate processing stages. In this case, made and bought parts must be collected together into assembly sets before assembly can start at the next stage.

There are, however, some simple products for which component processing and assembly can both be done at the same stage. Where this is possible it is usually desirable, because it reduces both throughput times and the stock investment.

An example can be found in a company making simple hand tools such as screwdrivers, wood chisels, and files. All these tools comprise a steel bit, a wooden handle, and a metal ferrule. The process routes for these tools start with a list of materials to be collected, including the bought ferrules and wooden handles. It then continues with the operations needed to make the steel tool, and to assemble it with the handle and ferrule, and it concludes with the last operation pack for dispatch.

Another example, where component processing and assembly can be combined, occurs in a cold-forge and welding group which flame-cuts parts from steel plate, and welds them together to form various boxes, cases and tanks. The process route for each item is:

(1) Give the specification and the size of the plate needed under material.
(2) Start with flame-cutting as the first operation, planned to produce product sets of parts from one or two different sizes of plate.
(3) Continue with machining, welding assembly, shot-blasting, heat treatment, and painting operations.
(4) End with a transfer to the assembly-department, say, group A3, for, say, product XXX.

One other item of information which should be included in process routes is the operation time per piece in machine-hours and/or in man-hours. This is needed by PC to check the load (in machine-hours and/or man-hours) against the capacity.

5.5 Parts lists

Parts lists, or bills of material as they are called in the USA, are needed for all complex assembled products. The need for a separate parts list can be avoided with simple assembled products by including this information in the process

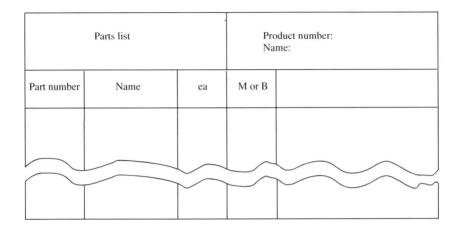

FIG. 5.7 A parts list. M = make, B = buy, ea = number per product.

routes, as explained in Section 5.4.2. In the general case, however, most assembly has to be done as a separate processing stage after component processing, and this requires a separate parts list for each product. The parts list should show for each part whether it was made in the factory, or if it was bought (M or B).

For convenience in PC, it is desirable that parts lists should be hierarchical in form, that is, they should be divided to show the parts for main assemblies, subassemblies, and sub-subassemblies and so on, as separate progressive sections in the parts list. This helps in the distribution of parts to assembly groups and to work stations inside those groups. A disadvantage of this arrangement is that some of the same part numbers will appear several times in the same parts list. For some types of PC, it may be necessary to have other consolidated lists of made parts and bought parts, where each part number, only appears once.

The information about the parts which should be included in the parts list is illustrated in Fig. 5.7. It comprises a part number a name, the quantity per assembly (ea), and the source (that is, whether it is made or bought).

5.6 Operation sheets

Operation sheets are normally only used for complex products. They show processing operations divided into a series of numbered work elements; they specify the tools needed for each element and they give the time needed to complete each element.

Figure 5.8 shows a pro forma for an operation sheet used for machining operations in a machine shop. It provides most of the information needed to set up the machine for an operation, including the identification number of the list of tools needed, the speeds and feeds to be used; and the time required for each element.

Operation sheet		Part number:		Operation number:		
		Part name:				
Machine number		Machine name:				
Tooling:						
Element		Time for each element (mins)	Speed	Feed	Tool-list number	
Number	Description					
01						
02						
03						
04						

FIG. 5.8 An operation sheet.

Operation sheets also provide the information needed for tooling analysis (TA). This subtechnique of PFA, is used in studies devoted to set-up-time reduction. Operation sheets also provide much of the information needed for methods study, aimed at the development of improved processing methods.

5.7 Lists of suppliers

If the call-off method of purchasing is used, (see Chapter 8) PC may be made responsible for calling-off periodic deliveries from suppliers (via a computer link). In this case, lists of suppliers and of the parts which they provide will be needed. There is currently, a tendency in some companies to combine production control and purchasing into one materials-management department. This brings together all tasks concerned with materials supply and purchasing into one department.

5.8 Lists of customers

In some companies, a large part of the output goes to a few major customers. Production control must know who they are and what products they buy, so that they can control the delivery schedules for these items.

5.9 Maintaining the accuracy of the database

There is an expression 'garbage in, garbage out' which points to the fact that if you feed incorrect data into a computer its output will be misleading rubbish. For consistent results in PC, it is essential that the database should be maintained at a very high level of accuracy.

One example will show the types of problem caused by inaccurate data. If it is found, in a factory, that the plant list is inaccurate and that some of the process routes show operations on machines which have been sold and which no longer exist, then any loading figures produced are rubbish. They will show the load on nonexistent machines, and they will understate the load on the machines to which the work has been transferred.

There are three main rules which help to maintain an accurate database:

1. There must be only one person with the authority to change data in the database.
2. There must be a strict method for authorizing and specifying only changes needed in the database.
3. There must be regular periodic systems for checking the accuracy of the database.

5.9.1 *Holding one person responsible for changes*

There must be only one database manager who is authorized to make changes in the database. This does not mean that this manager will necessarily plan or authorize the changes. It only means that this is the only manager who can make changes.

Experience has shown that where there are several people who can change a database it quickly becomes inaccurate, and it is impossible to determine who is to blame. With modern centralized data storage, it is not difficult to arrange that only one person can change the content of a database.

5.9.2 *Methods for introducing changes*

A list of the different types of PC data and of the departments in one company which are mainly concerned with planning and authorizing changes is illustrated in Fig. 5.9. There must be well-documented procedures for authorizing all such changes, and for introducing them into the database.

5.9.3. *Systematic checking of a database*

Systematic methods for authorizing changes in a database are not sufficient on their own to ensure accuracy. Regular systematic cross-checks are necessary to ensure the maintenance of a high level of accuracy. The following are some of the types of check which are used in practice:

Checking the plant list A physical check is made at the beginning of every quarter. Lists are made of the plant found in each department and group, and

Type of data	Department concerned	Standard procedure
1. Plant list	Buying	P1. New plant
2. Workers	Personnel	E1. Personnel change
3. Process routes	Product planning	M1. Method change
4. Parts list	Design	D1. Modifications
5. Operation sheets	Product planning	M2. Method changes
6. Suppliers	Buying	B1. Approved suppliers
7. Customers	Sales	–

FIG. 5.9 Authorized database changes.

these are cross-checked with the plant list. Any differences are investigated to find out why the database-change system failed to report them, and the database is corrected.

Checking the list of employees The numbers and identity of employees should be checked, say, four or five times a year, at random intervals. The aim should be to get together all the people in each group. It is not necessary to have special roll-calls. The checking can be done at briefing and training sessions. Any absentees should be visited during, or as soon as possible after, the meeting.

Process routes Three types of check are needed:

1. *Is there a process route for every part made?* The made parts in the parts lists should be cross-checked against the process routes. Any missing routes should be listed and replaced.

2. *Are all existing process routes necessary?* One way of making this check is to find any process routes for parts which have not been used in production or sold in (say) the previous two years. These parts are checked, and if it is confirmed that they are seldom or never needed them the routes are transferred to a *suspense file*, to save the cost of revising them every time old machine tools are sold, or new standards are introduced. In the unlikely event that these parts are needed again, the old routes must be revised before they are reused. It is not uncommon to find companies which believe they have 10 000 current parts, and to find on checking that the figure is nearer 2 500. It is important to base these checks on usage (for example, on sales) and not on the parts manufactured. It is not uncommon to find parts being made for which there can be no possible use. They may be made, for example, because a machine that is used to make them is short of work, or because a high rate of incentive bonus is paid.

3. *Are the routes accurate?* One check is to cross-check the routes against the corrected plant list, to find any routes which specify machines which are not in the plant list. These routes must then be corrected.

Checking the parts lists Checking that all the parts needed are included in the parts lists is partly self-checking. A product cannot be built without all the parts. The real dangers are: that a part may be listed in the parts list for only one product, but it may be needed for two or more products, and that there may be parts which are still in the parts list but which are no longer required.

The accuracy of a parts list can be checked by the carefully controlled issue of one set of parts against the parts list. The product is then built. Any shortages or surplus parts are listed, and the parts list is corrected.

Operation sheets Setters should be encouraged to check the operation sheet every time they set up for an operation and to report any errors for correction. The issue of tools to the setter should be checked against the operation sheet, and any errors should be reported.

5.10 Summary

The data needed for PC comprises: a plant list, a list of workers, process routes; product-parts lists; operation sheets, and, in some cases, lists of suppliers and customers.

Only the most complex types of industry need all this data. In the simpler types of production, only the first three data types are essential.

The quality and reliability of PC depend on the accuracy of the database. What is needed is the clear delegation of responsibility for making changes to one person, systematic and well-defined methods for planning and introducing changes, and frequent regular checking of the database for accuracy.

6

THE ANNUAL PROGRAMME FOR PBC

6.1 Introduction

Programming (called master scheduling in the USA) is the first level of production control (PC). It schedules the output of finished products from a factory. If standard products are produced, three main types of programme are required:

(1) *sales programmes*, which show products to be shipped to customers each period;

(2) *production programmes*, which show made parts and products to be completed each period;

(3) *stock programmes*, which show the stocks of finished products at the end of each period.

These programmes will show a list of all the products; a series of equal dated time intervals; and, at the points of intersection (for each type of programme) the number of products to be delivered to customers, or to be completed in the period, or to be held in stock at the end of the period (see Fig. 6.1).

If nonstandard products are produced (jobbing), it may only be possible to show broad categories of product types and the planned output may have to be measured in standard units (SUs) of length, area, volume, weight, etc., or in terms of money value (see Fig. 6.2.) In this case, only one production programme is needed. Because no stocks of jobbing products, can be held, the sales and production programmes will be the same, and no stock programme is possible.

The term of a programme (called the planning horizon in the USA) is the length of time it covers. Programmes with three different terms will usually be needed by a manufacturing company:

(1) *long term*, say five to ten years for corporate planning;

(2) *annual*, one year for financial, purchasing, and capacity planning, etc.;

(3) *short term*, using a standard period of, say one week for assembly control, for controlling the manufacture of products which are sold in component form, for controlling purchase deliveries with the call-off method, and as the basis for the control of component processing with Period Batch Control (PBC) in implosive, square, or explosive assembly industries.

Week number	1	2	3	4	5	6	7	8	9	10	11	12	13	14	15	16	17	18	19	20	21	22	23	24	25	26	Total
Product A Sales	10	10	10	15	15	15	15	15	20	20	20	20	20	20	25	25	25	25	25	30	30	30	25	25	25	25	540
Product A Prod.	15	15	15	15	15	15	15	15	15	15	20	20	25	25	25	25	25	25	25	25	25	25	25	25	25	25	540
Product A Stock	5	10	15	15	15	15	15	15	10	5	5	5	10	15	15	15	15	15	15	10	5	–	–	–	–	–	–
Product B Sales	25	20	20	20	20	15	15	15	15	15	15	15	15	15	10	10	10	10	10	10	10	10	10	10	10	10	360
Product B Prod.	25	25	25	20	20	15	15	15	15	15	15	15	15	15	10	10	10	10	10	10	10	10	10	10	10	10	370
Product B Stock	–	5	10	10	10	10	10	10	10	10	10	10	10	10	10	10	10	10	10	10	10	10	10	10	10	10	–

Week number	27	28	29	30	31	32	33	34	35	36	37	38	39	40	41	42	43	44	45	46	47	48	49	50	51	52	Total
Product A Sales	25	25	25	25	25	25	25	25	20	20	20	20	20	20	20	15	15	15	15	15	10	10	10	10	10	10	1015
Product A Prod.	25	25	25	25	25	25	25	25	20	20	20	20	20	20	20	15	15	15	15	15	15	15	15	15	15	–	1030
Product A Stock	–	–	–	–	–	–	–	–	–	–	–	–	–	–	–	–	–	–	–	–	5	10	15	20	25	15	–
Product B Sales	10	10	8	8	8	8	8	10	10	10	10	10	15	15	15	15	15	20	20	20	25	25	25	25	25	25	755
Product B Prod.	10	10	10	10	10	10	10	10	10	10	10	10	10	10	20	20	20	20	20	20	25	25	25	25	25	–	755
Product B Stock	10	10	12	14	16	18	20	20	20	20	20	20	15	10	15	20	25	25	25	25	25	25	25	25	25	–	–

Fig. 6.1 An annual programme for standard products.

Week number	1	2	3	4	5	6	7	8	9	10	11	12	13	14	15	16	17	18	19	20	21	22	23	24	25	26	Total
Sand slinger	15	15	15	15	20	20	20	20	20	20	20	20	20	20	20	20	20	20	20	20	20	20	20	20	20	20	500
Jolters	20	20	20	20	20	20	20	20	20	20	20	20	20	20	20	20	20	25	25	25	25	25	25	25	25	25	565
Manual	9	9	9	9	10	10	10	10	10	10	10	10	10	10	10	10	10	10	10	10	10	10	10	10	10	10	256
Total per week (tons)	44	44	44	44	50	50	50	50	50	50	50	50	50	50	50	50	50	55	55	55	55	55	55	55	55	55	–
Total per four weeks (tons)	–	–	–	176	–	–	–	200	–	–	–	200	–	–	–	200	–	–	–	215	–	–	–	220	–	–	1321

Week number	27	28	29	30	31	32	33	34	35	36	37	38	39	40	41	42	43	44	45	46	47	48	49	50	51	52	Total
Sand slinger	20	20	20	20	20	20	20	20	20	20	20	20	20	20	20	20	20	20	20	20	20	20	20	20	20	–	1000
Jolters	25	25	30	30	30	30	30	–	30	25	25	25	25	25	25	25	25	25	25	20	20	20	20	20	20	–	1165
Manual	10	10	10	10	10	10	10	10	10	10	10	10	10	10	10	10	10	10	10	10	10	10	10	10	10	–	506
Total per week (tons)	55	55	60	60	60	60	60	30	60	55	55	55	55	55	55	55	55	55	55	55	55	55	50	50	50	–	–
Total per four weeks (tons)	–	220	–	–	–	240	–	–	–	205	–	–	–	220	–	–	–	220	–	–	–	220	–	–	–	150	2671

FIG. 6.2 An annual programme for a jobbing foundry.

THE ANNUAL PROGRAMME FOR PBC 97

The long-term and annual programmes are forecasts, and because it is not possible to foretell the future, these programmes will normally require revision at varying intervals during the year. The short-term programmes will normally be based on existing sales orders, and, in any case, as they cover only a short term, and as they are issued at very short intervals, they should never need changing after issue.

This chapter looks at annual programmes only, and studies how they are planned.

6.2 Planning the annual sales programme

The periods used in constructing the long-term programmes need only be months, or even quarters. For the annual and short-term programmes, much shorter periods are needed, typically of one week. These shorter periods are essential for the PBC system, for both short-term programming and ordering, and they must also be used for the annual programme due to the need for the frequent reconciliation of achievement in meeting these two programmes.

The first task is to plan the annual sales programme for each product. There are four main tools available to help in making such plans:

(1) market research;
(2) business forecasting;
(3) statistical forecasting—trends;
(4) statistical forecasting—product life cycles.

Market research attempts to forecast future sales by addressing questionaires to people who are active in a particular market, such as major customers, wholesalers, retailers, and salespeople. *Business forecasting* uses changes in closely related published statistics, to indicate likely changes in sales. For example, Ministry of Agriculture statistics showing the number of farm acres put down to grass have been used to forecast future sales of hay-making machinery. *Statistical forecasting*, using a projection of past sales trends, is based on the idea that events tend to continue without change unless they are subjected to strong external pressures, such as new competitors entering the market, a price war, or new products introduced in the market which make a company's own products obsolete. *Statistical forecasting* based on trends is much less reliable today than it was fifty years ago, due to the greatly increased number of countries and companies in each market sector, and because of the big increase in the rate of changes in design.

An alternative or supporting method to forecasts based on trends is the assessment of the likely *product life cycle* for each product, based on recent experience, and the basing of sales forecasts on these cycles.

In planning annual programmes, the essential foundation uses statistical forecasting to find the sales programme. This should be supported by other methods, such as market research and business forecasting, as far as possible.

6.3 The method of statistical forecasting

Although statistical forecasting cannot foretell the future, it can, by basing its forecasts on the pattern of past events, find programmes which provide reasonable targets for future action. The idea that statistical forecasting is a method for forecasting the future should be forgotten, instead, the idea that statistical forecasting is a method for finding achievable targets for future production should be adopted. If this approach is adopted and all efforts are bent towards the achievement of the chosen targets, there is a reasonable possibility that the number of times that it will be necessary to revise the programme can at least be reduced.

Statistical forecasting starts with a record of past sales, such as that illustrated in Fig. 6.3(A). The effects of any abnormal events (such as a fire or a strike) must first be eliminated by revising the sales record to show the probable values for the sales if these events had not occurred (see Fig. 6.3(e)). It is then possible, using relatively simple statistical methods, to analyse this data to show:

(1) trends (are sales rising or falling?);

(2) seasonal variations;

(3) cyclical variations.

These constituents of the sales curve are also illustrated in Fig. 6.3.

6.4 Trends

Trends indicate the propensity of sales to grow or to decline. One method, known as the moving annual total (MAT) method, plots, at the end of each period, the totals of the products sold in the previous twelve months.

By looking only at figures for a whole year, this method eliminates all seasonal (and most random) variations, and it indicates only the rate of growth or decline in annual sales (see Fig. 6.4).

One can project from curves of MAT values, using the method of least squares, to indicate the probable sales for future periods of time (see Fig. 6.5). If the trend is projected on the basis of three different time periods (say twelve months, four months, and two months), three different forecasts will probably be obtained. In the case illustrated in Fig. 6.5, the fact that the twelve-month forecast shows growth while both the four-month and the two-month trends show a decline indicates that there may have been a market change, and that the twelve-month forecast is too optimistic. In this case, consideration of the normal life cycle for machines of the type under investigation, and the results of market research, gave grounds for the belief that some growth in sales was possible. The company selected a programme for sales for the following year which included a small increase over the previous year.

It will be seen that this method of choosing a sales programme, or a target value for future sales, is far from exact. The best that can be said for it is that

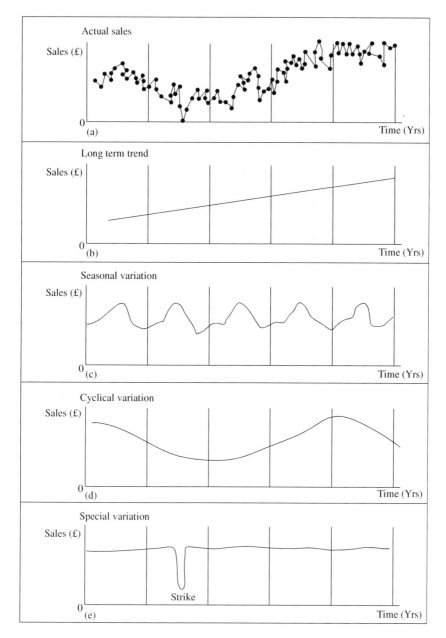

FIG. 6.3 Types of variation in a Sales curve: (a) actual sales, (b) a long-term trend, (c) seasonal variation, (d) cyclical variation, and (e) special variation.

it provides methods for ensuring that some of the more important factors in making a choice are systematically examined.

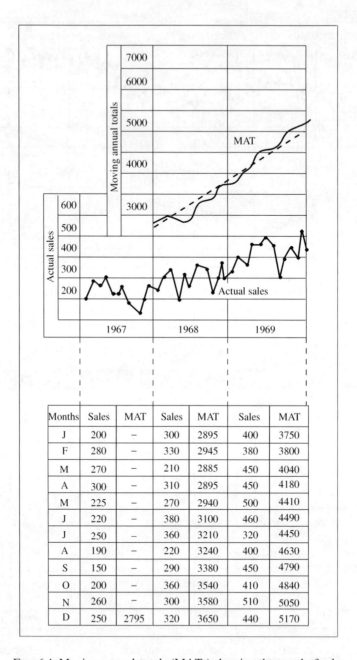

FIG. 6.4 Moving annual totals (MATs) showing the trend of sales.

Some would advocate the use of more sophisticated methods of statistical analysis. There are, however, so many variables involved in determining the

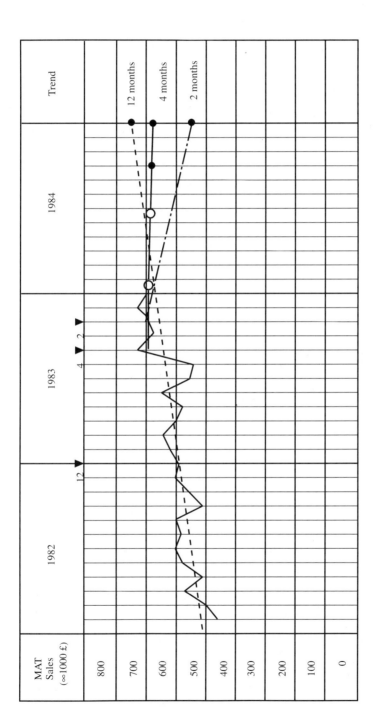

FIG. 6.5 A projection of MAT trends.

102 PROGRAMMING, ORDERING, AND DISPATCHING WITH PBC

actual level of the sales achieved that any forecast based on past results can only hope to be right within wide limits of accuracy. Under these conditions, it is difficult to justify the use of more sophisticated and expensive methods of analysis.

6.5 Seasonal variations

If one knows the actual sales for a product by period for two or more years, one can determine the percentage of the annual sales which is normally sold in each period. Figure 6.6 shows such a record.

The average sales per period and per year can be found by adding the period figures, and also the annual totals sold during previous years, and dividing them by the number of years. The percentage of annual sales which are sold in each period can be calculated. Applying these percentages to the chosen sales target for the next year makes it possible to find a reasonable distribution of the total between the periods in that year.

These seasonal variations depend on the general type of product and on the nature of the market where it is sold. Even with a new product, it is possible

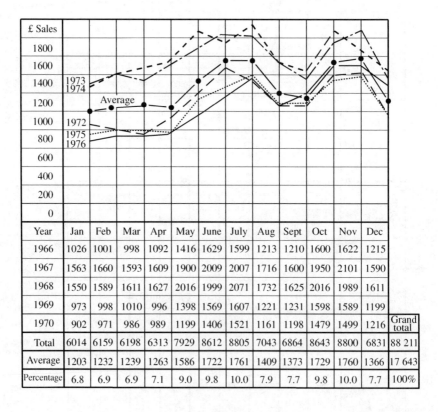

Fig. 6.6 The seasonal sales variation.

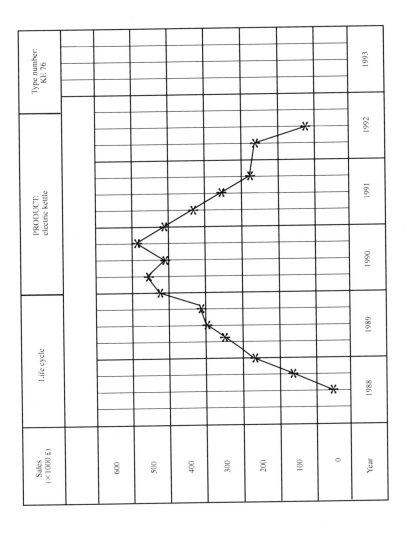

FIG. 6.7 A product life cycle.

to predict probable seasonal variations by basing them on other roughly similar types of product.

6.6 Product life cycle

The *life cycle* of a product is a historical record of sales between the date of its introduction and it's removal from the product range. Such a record is illustrated in Fig. 6.7.

Historically, there has been a dramatic reduction in the life of most industrial products during the past one hundred years. This reduction in life cycles has been caused by big increases in the number of different companies in the world producing most types of product. Each of these companies tries to obtain a competitive advantage by introducing new, improved products. This makes frequent changes in design necessary if a major company wishes to stay in the world market. The introduction of new models, or variants of an old product, may help to maintain sales for a time, but eventually some new product will reach the market which will make the old designs of product obsolete. When this point is reached, a company needs to have its own replacement product ready for launching.

To help in strategic planning, companies need to maintain records of the life cycles for all their products, and also, as far as possible, of the life cycles of their competitors' products.

6.7 Cyclical variations

In addition to the random, seasonal, and life-cycle variations, total sales will tend to rise and fall over a longer period. This type of change is called the cyclical variation. It is much less regular than the seasonal variation, and it is therefore much more difficult to predict. The life-cycle variation is normally concerned with major product models, whereas the cyclical variation studies the much longer life of a product range containing a series of models. It is often difficult to differentiate between these two types of variation.

It is usually possible to calculate, however, the past cyclical variations from past sales figures, as indicated by Fig. 6.3. The curve is never as simple as that shown in Fig. 6.3, and its amplitude and wavelength will be different for different types of product and in different historical periods. Since it is difficult to predict future changes in the cyclical-variation curve, one is obliged to try and steer to follow these variations as they occur.

The cyclical variation is concerned with both individual product sales and with total sales. It is influenced partly by the general state of trade in the country, by conditions in the industry of which it forms a part, by the standard of management in the company, and by the life cycles of the different products being produced.

The cyclical variation tends to peak at different times for different industries. The economic cycle in a country, varying between a boom and a slump, is at

THE ANNUAL PROGRAMME FOR PBC

(a) Seasonal sales, with make-to-order.

Month	J	F	M	A	M	J	J	A	S	O	N	D
Sales, product A							50	100	100	100	100	50
Production	?	?	?	?	?	?	50	100	100	100	100	50
Stocks							–	–	–	–	–	–

No production for six months.

(b) Spread sales, with a seasonal price, discount and credit.

Month	J	F	M	A	M	J	J	A	S	O	N	D
Sales, product A			20	20	30	30	50	50	50	100	100	50
Production	?	?	20	20	30	30	50	50	100	100	100	50
Stocks	–	–	–	–	–	–	–	–	–	–	–	–

No production for two months, and a wide variation for the rest of the year.

(c) Spread production with seasonal labour and overtime.

Month	J	F	M	A	M	J	J	A	S	O	N	D
Sales, product A			20	20	30	30	50	50	50	100	100	50
Production	30	30	30	30	30	30	30	30	30	80	100	50
Stocks	30	60	70	80	80	80	60	40	20	–	–	–

Heavy stocks and an uneven production rate.

(d) Export product to southern hemisphere countries, for example, Australia, New Zealand, South Africa, Argentina.

Month	J	F	M	A	M	J	J	A	S	O	N	D
Sales, product A, UK							50	100	100	100	100	50
Sales, S hemisphere	50	50	50	50	50							50
Production	50	50	50	50	50	50	50	50	100	100	100	100
Stocks	–	–	–	–	–	50	50	–	–	–	–	–

(e) Introduce a seasonally balanced product range. In this example all the products have the same stock value per product.

Month	J	F	M	A	M	J	J	A	S	O	N	D
Sales												
Product A							50	100	100	100	100	50
Product B	100	100	50									50
Product C			50	100	100	100	50					
Production	100	100	100	100	100	100	100	100	100	100	100	100
Stocks	–	–	–	–	–	–	–	–	–	–	–	–

FIG. 6.8 A seasonally balanced product range.

least partly caused by the peaks and troughs of the cyclical variations in different industries drifting into and out of phase.

6.8 Programme reconciliation

In the next chapter, we will study flexible programming, based on a series of short-term programmes. Because it not possible to foretell the future, it is inevitable that the cumulative totals for the sales in each period which are forecast in the annual programme will tend to vary from the cumulative totals indicated by the series of short-term sales programmes, which are based mainly on actual sales (see Fig. 3.5).

When this happens, it is necessary to revise the annual programme, and also to revise any budgets, purchase contracts, or other plans which have been based on the annual programme. This reconciliation has already been considered in Chapter 3.

6.9 The annual production programme

The annual production programme is only used for standard products. It is derived from the annual sales programme. The objectives in planning this programme are to provide products at the times needed to achieve the sales programme, and at the same time to maintain an even rate of production with a minimum investment in stocks and a minimum loss in capacity.

Figure 6.8 illustrates the problem which arises if sales are highly seasonal. If products are only produced when they can be sold, there will be major losses in unused factory capacity, in some periods of the year. If, on the other hand, products are produced at an even rate throughout the year, there will be a serious increase in stocks and in capital tie-up during the low-demand season.

A small amount of *smoothing* is possible by offering discounts, or long credit for out-of-season sales, and by working overtime in the peak-demand period and working short time during the dead season, or, for some products, by developing sales markets in the southern hemisphere, where the seasons are reversed.

The only permanent way to overcome a highly seasonal sales market, however, is to develop a seasonally balanced product range. As an example, one company in the agricultural-machinery industry, which specialized in grass-cutting and hay-making machinery, balanced their product range seasonally by the addition of ploughs, hoes, disc harrows, and seed drills to the range which could be made and sold in other periods of the year.

6.10 The finished-product stock programme

A finished-product stock programme, which is required with standard products, is found by simple calculation from the sales and production programmes. An example was given by Fig. 6.1.

6.11 Summary

Annual programmes are needed in industry as the basis for medium-term plans, for example, financial budgetary control, cash flow, purchase contracts, and other medium-term needs.

Annual programmes for standard products are based on the number of products to be produced. Three related programmes are generally needed covering sales, production, and finished-product stocks. For jobbing products, it is impossible to hold stocks and only one programme is needed. The output in this case is shown in standard units (SUs), or in units of money value.

Annual programming starts with an annual sales programme or forecast. Market research, business forecasting, and/or statistical forecasting can be used to plan realistic sales programmes (targets). With standard products, attempts are generally made to *smooth* the production in order to conserve capacity. This requires the preparation of *production* and *finished-product stock* programmes, in addition to the sales programme.

A major problem in programming arises when sales markets are highly seasonal. The best solution for this problem is to develop a seasonally balanced product range.

7

SHORT-TERM OR FLEXIBLE PROGRAMMING FOR PBC

7.1 Introduction

Flexible, or short-term programming is the name given to programming which is based on a series of equal short terms or time periods. It is called flexible programming because the use of short-term period programmes makes it possible to follow variations in market demand without an excessive accumulation of stocks, without obsolescence, and without the need to change programmes and related orders once they have been approved and issued. Such changes are inefficient because they tend to leave remnants of part-finished work in the factory.

Flexible programming is the first level of production control (PC) special to Period Batch Control (PBC). At this level, the standard *period* and the standard *programming schedule* are fixed. Period batch control is a *just-in-time* (JIT) *system*, and one of its main aim is the minimization of stocks. It is recognized, however, that in some cases stock may be needed to either conserve capacity (a smoothing buffer), or to make selling possible in ex-stock markets (delivery buffer), or even in the early stages of introduction, as (insurance buffer) against the risks of scrap or of late delivery by suppliers. The long-term objective is to eliminate buffer stocks. The short-term objective may be to achieve this aim with the minimum of buffer stock.

For standard products, as in the case of annual programmes, there is again a need in short-term programmes for sales, production, and finished-product-stock versions of the programme. This chapter studies how flexible programmes are planned, issued, and controlled.

Three main cases are examined:

(1) standard assembled products;
(2) standard piece-part products;
(3) jobbing products.

These cases are followed by a study of the case of short-term programming in the simple *process industries* (see Fig. 1.3).

Standard products are products made to a predetermined designs, either continuously or intermittently in a succession of batches. They may be either assembled products, made by bringing together a number of different component parts (see the explosive industries in Fig. 1.3), or they may be piece-part products (see the implosive and square industries in Fig. 1.3), or they may be bulk-material products made in the process industries (see again Fig. 1.3).

Jobbing products are products which are special to one particular order. Occasionally, such orders are repeated, but there is no way of knowing if this will happen when the first order is received, and there is no possibility of manufacturing for stock. Jobbing products may be assemblies or piece parts. They may be ordered as single items, or as a batch.

7.2 Selecting the programming period

Period Batch Control requires the selection of a single standard scheduling *period* for all production. It is convenient if the time chosen for the standard period is based on equal well-established time intervals, such as a week, a day, or an hour. Months should not be used because they vary in the number of days per month, but four week periods (thirteen per year) are possible.

The objective is always the same, to minimize the stock investment by finding the shortest period which is long enough for the completion of any part or assembly made in the factory. In other words, all made items must have a throughput time which is less than the chosen period. In practice, if Group Technology (GT) is used—which greatly reduces throughput times—it is seldom necessary to use periods longer than one week. This allows a maximum throughput time per period (week) of 168 hours. There are very few manufactured components which cannot be completed in less than 168 hours.

Exceptionally, longer periods may be needed in industries where ageing, or maturing of the product, or the growing of cultures is part of the process, and shorter periods may be required in industries, such as food manufacture, where the shelf life is a significant factor.

In process, implosive, and square industries, which make piece-part products, and which regulate their manufacture by the issue of production programmes, the choice of the period depends solely on the needs of programming. In explosive industries, consideration must be given to the throughput times for assembly and for batches of made parts, which must be manufactured for assembly, and also to the lead times required for the call-off of purchases from suppliers.

7.3 Selecting the programming schedule

Period Batch Control requires the adoption of a single standard programming schedule which is repeated every cycle. Once again, the objective is to minimize the stock investment by finding the shortest schedule and period—throughput time—which can be used in practice.

A number of examples of programming schedules from practice are illustrated in Fig. 7.1. The objective, as stated above, is to minimize the number of

110 PROGRAMMING, ORDERING, AND DISPATCHING WITH PBC

(a) A two-period schedule

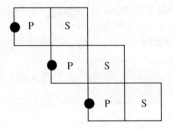

Key:
● = Programme meeting
P = Production programme
S = Sales programme
Ac = Accumulate sales orders
M = Obtain materials

Note, the orders accumulated in week 1 (Ac) form the sales programme (S)

(b) A three-period schedule

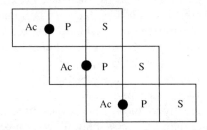

(c) A four-period schedule

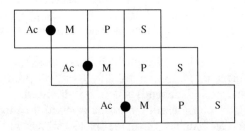

FIG. 7.1 Examples of programme schedules.

stages in the schedule. The ideal would be one single stage, but this is not generally possible. The reasons necessitating the addition of stages to a programming schedule include:

(1) The addition of an accumulation stage for the accumulation of orders into batches before processing;
(2) The addition of a stage for obtaining materials for processing;
(3) The addition of a stage to allow additional time for processing.

In process, implosive, and square industries, where the products are generally piece-parts or bulk materials, the programming schedule found at this stage is used as the foundation for Period Batch Control. In the explosive industries, the programming schedule must be expanded to provide an ordering schedule, to include time for component processing to make the parts needed for assembly. This modification of the production schedule is examined in the next chapter.

7.4 Standard assembled products and their programming schedule

For the programming schedule for standard assembled products, four general cases will be considered:

(1) assembled products with few varieties, low demand variation, and short throughput times;
(2) the same but with long throughput times;
(3) assembled products, with many product varieties and/or high demand variation requiring *smoothing*;
(4) assembled products which can only be sold if immediately available from stock.

Each of these cases will be examined in turn.

7.4.1 *Assembled products with few varieties, low demand variation, and short throughput times*

An example might be the assembly of a range of centrifugal pumps (illustrated in Fig. 7.2). One-week periods have been selected together with a programming schedule of three weeks per cycle, in which sales orders accumulated in week 1 and they are assembled in week 2 for delivery to customers in week 3, at the end of each three-week cycle.

Apart from approving the sales and assembly programmes, the programme meeting in this case (at the end of week 1 in each cycle) will be concerned with checking the programme load (in man-hours) against the available capacity and with changing the capacity level when necessary; secondly, it also reconciles the cumulative output planned in the annual programme with that in the series of short-term programmes, and it modifies the annual programme when this becomes necessary, to bring the two into line.

Copies of the forms used for the period sales and production programmes, approved at the programme meeting in the centrifugal-pump case, are illustrated in Fig. 7.2(b, c). The load imposed by the production programme on the available capacity must be checked each week before issuing the programme. If the trend of the load is rising or falling, it will eventually be necessary to change the capacity level. In the case of the centrifugal pumps, both the load and the capacity were measured in assembly man-hours, and the capacity was changed in the short term by such methods as overtime and the employment of temporary contract labour.

(a) A flexible-programming system

Week number	1	2	3	4	5
Cycle 1	Ac●	P	S		
Cycle 2		Ac●	P	S	

Key:
Ac = Accumulate
P = Production
S = Sales
● = Programme meeting

(b) A period sales programme (S) found by accumulating orders received in week 1

Period sales programme (pumps)				Week number 3
Code	Description	Size	Quantity	

(c) A period production (assembly) programme (P) found by smoothing the sales programme

Period production programme (pumps)					Week number 2
Code	Description	Size	Quantity	Man hours ea	Total

FIG. 7.2 Short-term programmes for a range of centrifugal pumps.

Figure 7.3 shows the records considered by the programme meeting when the pump programme is considered. It shows, first, the record of actual sales and of the trend moving annual totals (MAT) of orders received in the previous five weeks. This is used to indicate, if sales are rising or falling. The second record (Fig. 7.3(b)) shows the actual load per week, compared with the present capacity level. The additional capacity given by 5, 10, and 15 hours per week overtime and by employing one additional assembly worker is also known. Armed with this information, the programme meeting decides when a change in capacity is needed, and how to achieve it. The third record illustrated in (Fig. 7.3(c)) shows how the cumulative output totals in the series of

SHORT-TERM OR FLEXIBLE PROGRAMMING FOR PBC 113

(a) Sales orders received

Product				Week number			Date	
Type	Name	Quantity	MAT Week	MAT Week	MAT Week	MAT Week	MAT Week	

(b) The load and capacity

Load and capacity			Group number	Week number	Date		
Machine			Capacity	Load	+	−	
Type	Plant number	Name					

(c) Reconciliation with the annual programme

Reconciliation with the annual and short-term programmes		Product			Year		
Week number	Date	Short-term programme		Annual programme		+	−
		For the week	Cumulative	For the week	Cumulative		

NB + and − indicate the difference between the short-term cumulative total and the annual programme cumulative total; underlined if they are short-term cumulative total is the larger value.

FIG. 7.3 The records for a programme meeting.

short-term programmes compares with the same totals from the annual programme. This information allows the programme meeting to decide when it is necessary to modify the annual programme.

7.4.2 *Assembled products with few varieties, low-demand variation, and long throughput times*

In the example of pump assembly, the throughput time for one pump was sixty minutes. The first pump was finished sixty minutes after starting work on a

Monday morning, and the last assembly for the week had to be started not less than sixty minutes before ending work on a Friday afternoon.

The choice of the standard period and of the programming schedule depends on the assembly throughput time. Figure 7.4(a) shows the case of a tractor digger for which the throughput time could not easily be reduced below six days. One week periods were used. Two weeks were allowed in the programming schedule per product for assembly, as follows:

(1) week 2 and part of week 3, weld the frame, stress relieve, shot blast, paint, and machine the frame;
(2) week 3 and part of week 4, final assembly.

(a) A long throughput time of six days per digger (3+3)

Cycle	Week number					
	1	2 MTWTFSS	3	4	5	6
1	Ac	● P1	P2	S		
2		Ac	● P1	P2	S	
3			Ac	● P1	P2	S

Key: Ac = accumulate orders, P = production programme,
P1 = frames, P2 = final assembly, S = sales programme

(b) A shorter throughput time of four days per digger (2+2)

Cycle	Week number					
	1	2	3	4	5	6
1	Ac	● P	S			
2		Ac	● P	S		
3			Ac	● P	S	

P = P1 + P2

FIG. 7.4 A programme schedule for a tractor digger.

The first digger-frame welded assembly, for each cycle, is started on Monday morning in week 2 and it is finished by Wednesday evening in the same week. The last digger-frame is started on Friday in week 2, and it is finished by Wednesday evening in the following week, week 3.

The first main assembly in the cycle (P2) starts on Monday morning in week 3, and it is finished by Wednesday evening in week 3. The last digger in the main assembly in each cycle starts on Friday in week 3, and it is finished on Wednesday evening in week 4.

Delivery to customers can start on Thursday in week 3 with the completion of the first digger, and the remaining deliveries must be completed by the end of week 4. It should be noted that a two-day (Thursday and Friday) time buffer has been allowed between frame assembly and main assembly. The first frame is ready for main assembly on Thursday morning each week, but the main assembly is not scheduled to start until the following Monday.

Figure 7.4(b) shows that if the throughput time for these two stages of make frame and welding, plus final assembly, can be reduced from six days to four days, it will then be possible to reduce the production schedule from four weeks to three weeks. As an alternative, this saving can be made by eliminating the time buffer and allowing an increased overlap between the P and the S stages.

Period-Batch-Control purists have criticized this use of overlapping schedules on the grounds that it complicates the allocation of responsibility to individuals for completion by the due date. It was decided that the major reduction in the stock level more than compensated for some additional complexity in system control.

7.4.3 *Standard assembled products with many varieties and/or high demand variation*

An example in this case might be a production programme for the assembly of a wide range of taps, mixer valves, and other plumber's fittings for domestic kitchens and bathrooms.

The high demand variation in this instance, will probably make it necessary to smooth production, making additional products of the most popular sales varieties for stock, when sales orders are below the mean-demand-capacity level, and using these stocks to supplement production when sales orders are above the mean-demand-capacity level.

One-week periods and a three-week programming schedule have again been selected in Fig. 7.5, which shows both the regular cycle of accululating orders, assembling and delivering to them customers, and also shows the mechanism of the smoothing system. Figure 4.7 shows smoothing at a fixed rate of sales in another factory, and Fig. 7.6 shows as an example, how smoothing can handle changes in sales demand which require changes in capacity.

The weekly programme meeting checks and authorizes the issue of the production (assembly) programme and of the sales programme. It also watches the trend of orders received, adjusts capacity levels when necessary, and it

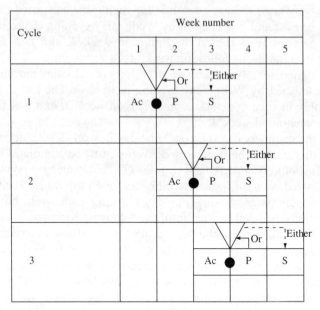

Note, P is at a fixed rate. If P is greater than S, the surplus goes to smoothing buffer stock. If S is greater than P, the shortfall is made up from the buffer stock.

Key: Ac = accumulate orders ● = Programme meeting
P = Assembly   S = Sales programme
▽ = Smoothing buffer stock

FIG. 7.5 A short-term programme with smoothing.

reconciles the cumulative output shown for the end of each period in the annual programme with that given by the series of short-term programmes.

The main aim of smoothing is to save capacity, which under conditions of high sales variability might otherwise be lost. If a company is operating well below its capacity level, smoothing may be unnecessary and uneconomical. If, however, a company is operating at or near the limit of its capacity, smoothing can be much more economical than buying additional machine tools.

There may be advantages if any smoothing stocks of finished machines are stored in the assembly group which makes them. This group should make a weekly report—giving the quantities of each product held as smoothing stock—to the secretary of the weekly programme meeting.

7.4.4 *Standard assembled products which can only be sold from stock*

There are some assembled products which can only be sold from stock. If a manufacturer cannot deliver from stock in these cases, the order will be lost.

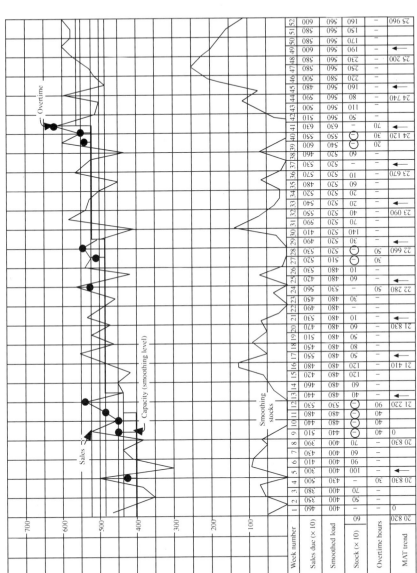

FIG. 7.6 Capacity regulation with growing demand. Note: (1) if the stock equals zero for two consecutive weeks, one capacity unit should be added; (2) if the sales are below capacity for five consecutive weeks, the capacity should be reduced by one unit. — = sales greater than capacity; ⊖ = sales greater than capacity for 2 weeks running; → = trend increasing.

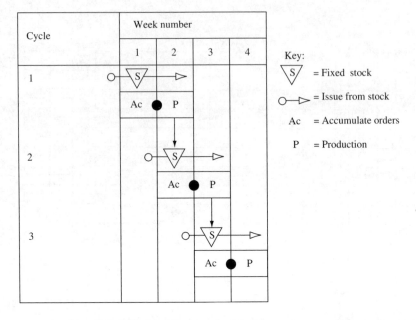

FIG. 7.7 Period batch control with delivery from stock.

This may be due to custom in the trade, or simply because other suppliers are willing to offer this service, and it is valued by customers more than small reductions in selling price.

Under these circumstances, flexible programming needs to find a system which provides delivery from stock with the minimum investment in stock. Figure 7.7 shows the general solution of this problem with Period Batch Control (PBC). A standard delivery-buffer stock level is fixed for each product. The goods are taken from stock as the orders are received, and then are immediately delivered to the customers. The quantities ordered and delivered to customers in each period are accumulated, listed, and they are ordered at the end of the period, for manufacture in the following period. As these products are completed, they are returned to stock to replace those issued previously to customers, in order to maintain the standard delivery-buffer stock level.

The stock required to give reliable delivery ex stock in this case, will be less than a maximum demand rate or two weeks. Some additional smoothing stock may also be needed. An intelligent estimate of the buffer stock level is all that is needed at the start. If after (say) ten weeks, the stock has never dropped below eight days average supply then the buffer should be reduced by (say) four-days supply. Fixing the starting value is not important. What is important is a control which regularly adjusts the buffer.

7.5 Products sold as piece-parts

We first looked at short-term (flexible) programming for standard assembled products made in the explosive industries (see Fig. 1.3), we must now examine the flexible programming needs for the implosive and square industries (see Fig. 1.3) where the products take the form of parts (piece-part products).

A later chapter on PBC ordering, Chapter 8, will study the problem of ordering firstly components for assembly (both bought and made), and secondly of ordering the materials for made parts. In the case examined here, the first of these ordering needs is not required because there is no assembly and the programme itself provides the list of parts to be made. The second ordering need, the ordering of materials, is, however, still necessary.

7.5.1 *Cast-iron foundries*

A foundry is used as an example of piece-part production, making castings—mainly on call-off—for a number of customers in the engineering industry, who machine parts and assemble engineering products.

The processes carried out in a foundry are:

- (1) metal melting;
- (2) laboratory-metal test;
- (3) sand treatment;
- (4) moulding;
- (5) core-making;
- (6) pouring metal;
- (7) cooling;
- (8) knocking-out;
- (9) fettling;
- (10) shot-blasting;
- (11) painting;
- (12) storing and shipping.

The productive processes of moulding, core-making, pouring, cooling knocking out, and fettling are divided between three main GT groups:

Group G1, the sand-slinger group.

Group G2, the jolt-moulding group.

Group G3, the manual-moulding group.

They are supported by three service groups:

Group S1, metal melting (includes laboratory and sand treatment).

Group S2, painting (with shot-blasting).

Group S3, shipping (with finished-product storage) (see Figure 7.8(a)).

To make it possible to deliver against short call-off lead times, some delivery-buffer stocks are held of all the castings made; call-off orders are delivered from stock to customers on receipt of regular periodic call-off instructions. Orders to replace these issues are then given to the foundry in the form of short-term programmes based on the accumulation of the orders received.

The flexible-programming method used is illustrated in Fig. 7.8(c). Periods of one week have again been selected, with a three-week schedule. Castings are

120   PROGRAMMING, ORDERING, AND DISPATCHING WITH PBC

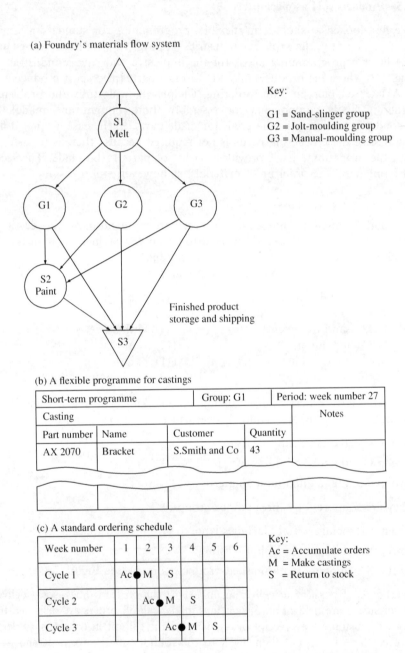

FIG. 7.8 Period batch control for a cast-iron foundry.

delivered to customers, and call-off orders are accumulated in week 1. The quantities in stock are subtracted from the accumulated orders at the end of the week to find the *must-make quantity*. There are wide variations in the number of castings ordered per week, and smoothing is necessary to reduce capacity losses. If the total must-make quantity is less than the manufacturing capacity (in tons of castings per week), additional quantities are made of the most popular items, that is, of those castings which are normally ordered by some customers every week.

The castings are made in week 2, and they are sent upon completion to stores, to replace those delivered from stock to customers. Similar systems to this can be used, with or without delivery-buffer stocks, in potteries, glass works, and in other implosive industries, which make standard piece-part products.

7.5.2 *Deliveries of standard implosive products from stock*

It will be realized that delivery from stock is only possible for standard products. It is possible to use the method already illustrated for assembled products in Fig. 7.7 to control sales from stock of piece-part products made in implosive industries. It is, however, much more difficult in this case to fix a stock level for each part which will give a reliable delivery from stock, and the total stock investment needed can be high.

If it is decided to deliver implosive products from stock, the standard stock level to be held of each item must be fixed. Sales orders will be delivered from stock as soon as they are received (as in Fig. 7.8). The orders will be accumulated and listed in the first period of each cycle, and the items in these accumulation lists will be ordered at the end of the period, for manufacture in the following period. As each item is completed it will be sent to the store to replace items previously sent to customers. Once again, fixing the starting value for the buffer is less important than the need to control its level in operation.

7.6 Jobbing products

Jobbing products are special products (nonstandard) which are normally ordered in small quantities, and for which it cannot be known when an order is received if that order will ever be repeated. Jobbing products may be made from standard materials using standard tools, but even in these cases there is a high probability that any jobbing order will include some product design and/or production-planning tasks, to make the order ready for manufacture.

Examples of these *make-ready* tasks include the following:

(1) make a detailed drawing of a part;
(2) design a special pattern sheet for a decorative laminate;
(3) prepare a process route;
(4) select and buy the materials needed;

(5) order a pattern for casting;
(6) order dies for a forging.
(7) order any special tools needed, for example, form tools, taps, thread dies, hobs, etc.
(8) order special paint.

The difficulty with tasks such as these is that they are likely to have widely differing lead times. These lead times have to be estimated when quoting and making a delivery promise to a customer, but they cannot easily be assigned to a standard scheduling period for PBC.

The general solution in this case is illustrated in Fig. 7.9. Make-ready is treated as separate from manufacture and it is scheduled independently. When the make-ready tasks have been completed, the order is transferred to a ready file. Orders are selected from the ready file at the programme meeting for inclusion in the production programme for the next period. To permit smoothing and to use the available capacity to the best advantage, it is desirable that there should always be more than one-weeks load in the ready file.

7.7 Short-term programming for process industries

Process industries, as shown in Fig. 1.3, are simple industries in which a small number of materials are converted into a small number of products by a relatively standard succession of a small number of processes.

To be competitive, such industries must offer short, or ex-stock delivery. This process-industry category includes a large part of the food industry, where the shelf-life is a major restriction on the allowable throughput times. For these reasons, the standard period for programming in process industries is generally a day or even a shift.

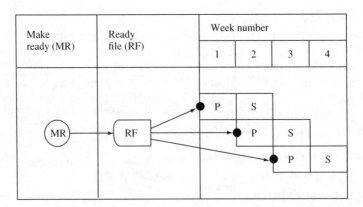

FIG. 7.9 Period batch control for jobbing production.

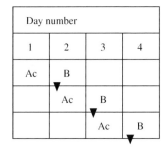

Key:
B = Bake bread and deliver it to the customers
Ac = Collect orders for the next day during the delivery round
▼ = Start the delivery round

FIG. 7.10 Period batch control in a small bakery.

Figure 7.10 illustrates the case of a small bakery making bread in a country town. The company has a shop in the high street, and deliveries are also made daily to local shops, hotels, restaurants, schools, other institutions, and some private homes. The delivery people both deliver and collect orders for the next day.

The standard period is one working day (Monday to Saturday), and the following standard schedule is two days: (1) accumulate orders from delivery vans, etc.; (2) make the bread (on an early morning shift) and deliver it.

7.8 The programme meeting

An important element in most flexible programming systems is the programme meeting. The chairman of this meeting has the following responsibilities:

(1) authorization of the issue of the short-term sales and production programmes;
(2) authorization of changes in manufacturing capacity, when they are needed;
(3) authorization of revision of the annual programme when reconciliation shows that it is needed;
(4) keeping the board of directors in touch with sales trends, and with their effect on manufacturing.

It is desirable that the meeting should be chaired by a director—normally the production director—who, as a member of the board of directors, can keep them in touch with manufacturing trends and problems. Other members will normally include the sales and production managers and the head of PC.

A programme-meeting secretary should be appointed, to prepare the information needed for the meetings. This involves the following tasks:

(1) *a capacity check*, where the volume of sales orders received (load) is charted and compared with the established capacity level to indicate if a change in capacity is needed;

(2) *programme reconciliation*, where the cumulative value of sales per product in the annual sales programme is checked against the same figure for the series of short term-programmes to indicate if a change is needed in the annual programme;

(3) *smoothing*, where the load imposed by the accumulation of orders for the period is checked against the established capacity level to determine if additional smoothing stock should be ordered;

(4) *sales and production programmes*, where draft programmes are prepared.

Simple computer software can be used to prepare this information very quickly, just before the meeting. Figure 7.3 has already shown a typical set of information sheets prepared for a programme meeting. The meeting itself will normally be short, with a fixed agenda on the following lines:

1. Consider the choice and quantity of *smoothing* products to be made (if any), and approve specific quantities.
2. Consider the draft *period sales and production programmes*, and revise and/or approve them.
3. Consider the relationship between *load and capacity*, and change the capacity level if necessary.
4. Consider the *programme reconciliation*, and amend the annual programme if necessary.
5. Any other business.

7.9 Summary

Nearly every factory in manufacturing is unique. To operate efficiently, a factory needs a combination of features in its operating system, not all of which will be required in any other factory.

The cases used in this chapter to illustrate different types of flexible programming only illustrate the range of possibilities. Any particular factory may have similar problems to those described for one of these cases. On the other hand, it may be concerned with several problems at the same time, making, for example, both standard products (some being sold from stock) and also jobbing products. Another problem is that products and markets change with time, and it may be necessary to modify the PC system to accommodate such changes.

This chapter has studied the short-term programming system for PBC. If it is decided, in the case of explosive industries, to expand the PBC system to cover parts manufacture, it will be necessary to modify the standard schedule (as explained in the next chapter).

One of the advantages of PBC, is that one simple PC system, based on a series of equal time periods and on a fixed schedule per cycle, can be used, with minor variations, to regulate the production of a wide range of different types of product.

8

ORDERING AND PURCHASING WITH PBC

8.1 Introduction

In the simple process industries, continuous line flow (CLF) is the normal type of materials-flow system, and the ordering of the small number of material items needed can be based on the frequently monitored trend of past usage. Because large amounts of one or a few varieties of material are required, statistical forecasting is a reasonably accurate method for determining short-term future needs.

In the implosive industries, the products produced are mainly parts which are listed for manufacture in a series of short-term programmes based on sales orders. The only other ordering required is for the small number of different material items used to make the products. Examples include pig iron, scrap iron, and sand for a foundry, and clay and glazing materials for a pottery. Once again, the ordering of these materials can be safely based on the trend of past usage.

In the square industries, the products are components, and in this case the materials are often those same components before treatment. They are usually provided by the customer with his order. Examples are bolts of cloth for dyeing in a dye mill, or for finishing in a finishing mill, or parts for an X-ray examination. The only materials to be ordered are process materials such as the dyes and X-ray plates.

It is only in the explosive industries that the completion of assemblies is scheduled by programming; and that ordering is concerned with the provision of: bought and/or made parts for assembled products, with the ordering of the materials needed to make made parts; and with the ordering of any special processing materials, such as coolants, welding rods, chills, etc.

This chapter starts by studying the purchasing function which procures most of the materials used in industry; then the simple systems needed for ordering materials in the process, implosive, and square industries are considered; and finally the problem of ordering bought materials and component parts is examined for assembly in the explosive industries.

8.2 Purchasing

Until recently, the most common type of purchasing for manufacturing was *batch buying*. Items were ordered in specified batch quantities for delivery by specified due dates. This type of buying is based on the outmoded idea, inherited from generations of merchants, that the best way to maintain a reliable supply of an item, is to maintain a stock of the item in a store.

Batch buying was closely associated with the method of stock control, which uses reorder levels in a declining stock to trigger the release of new orders, and with the pseudoscientific theorem of the economic batch quantity (EBQ)—called the economic lot size in the USA (see Fig. 1.9). In practice, this combination of ideas leads to a very heavy investment in stocks and to instability caused by the system's dynamic problems. The stock control ordering method is probably obsolete now, at least for regulating materials supply for manufacturing.

The scheduling of purchase deliveries, based on explosion from long-term (annual) programmes, has been widely used as an alternative, but it suffers from the general unreliability of long-term forecasts which require frequent revision of the schedules.

A more flexible and reliable alternative to batch buying is the call-off method. Typically, a long-term forecast of the probable requirement for an item is made by calculation (explosion), usually from an annual production programme. The supplier is given a purchase contract together with a copy of the forecast for information, but delivery against this contract may only be made on receipt of call-off instructions (call-off notes), issued at regular PBC period intervals.

To ensure reliable supplies of an item with the call-off method, it may be necessary for the supplier to hold some stocks. Whether this is necessary or not, will depend (as will the amount of stock needed) on the run quantities used by the supplier for manufacture; on his throughput time for manufacture, and on the call-off lead time.

If, for example, the call-off lead time is two weeks, the throughput time for manufacture is one week, the supplier has reserved capacity and is organized to manufacture in the exact quantities which are called-off each period, and the materials required can be obtained from stockists in less than one week, then the supplier need hold no stocks.

If the conditions are the same, except that the call-off lead time is one week, then the supplier must hold stocks to cover at least one week's maximum demand. Purchased items will then be delivered from stock on receipt of the customer's call-off instructions; materials will be obtained to cover the manufacture of this quantity in week 1; the parts will be made in week 2; and, on completion during the week, they will be returned to stores to replace the items issued from stock in the previous week (see Fig. 8.1.)

There has been a tendency in Britain to accept the need for close sales contact with customers, but to see purchasing as a function which can be managed by letter or by telephone. Because this has not resulted in good service, there has been a further tendency to attempt to reduce risks by dividing orders between a number of different suppliers.

This has not worked either, and some companies are now moving towards single sourcing for parts and to fewer sources (more parts per source). This

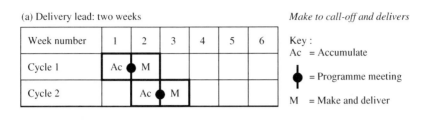

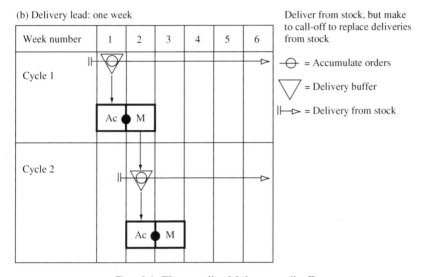

FIG. 8.1 The supplier Makes to call-off.

reduces the number of purchase suppliers, and the buyer can maintain much closer contact with this smaller number. In some cases, it has been the increase in business due to single sourcing, which has been the key to acceptance of the call-off method by suppliers.

The call-off method requires face-to-face negotiation with suppliers, and it takes a long time to change over completely. The benefits increase with each supplier who accepts the change. There may be a few suppliers who refuse to change at first. With persistence, they can be persuaded eventually, or they can be replaced.

8.3 Ordering materials for process and implosive industries

In both the process and the implosive industries, a supply must be maintained of a very small number of different basic materials. While it is difficult or impossible, particularly in the implosive industries, to forecast the sales of the many different components produced, it is much easier to forecast the requirement for the common materials used to make them.

The general solution to this ordering problem is to base ordering on the trend of past usage. Small buffer stocks are maintained of each item. A high delivery frequency is required to reduce the investment in stocks, and there is provision in the contract for varying the periodic delivery quantities when necessary, to prevent excessive growth or exhaustion of the buffer stocks.

Three examples from practice will show how this ordering system is used in different industries.

8.3.1 *Limestone for a cement works (a process industry)*

In this case, the limestone quarry is near the cement works, to which it is connected by a conveyor. Projection from the trend of past consumption is used to produce a running quarrying programme, which covers four months two months firm—will not be changed and two months reconsidered at regular monthly meetings before being made firm. These meeting add, each month, one new firm month to the programme. A small buffer stock is held in the quarry. The amount of limestone in this buffer stock is measured (estimated) each week. If it exceeds two-week supply, the quarrying programme may be reduced to reduce the stock level. If the buffer stock drops towards zero, it may be necessary to increase the quarrying programme. If the sales of cement rise or fall, the stock of finished cement will also change, requiring a change in the production rate. This will automatically change the quarrying programme.

8.3.2 *Pig iron for a foundry (an implosive industry)*

Pig iron is delivered to a foundry from a supplier's blast furnace, to a delivery schedule found by the continuous projection of past consumption. Deliveries are made by rail at an average rate of four truck loads per week. The volume, and thus the weight of the pig iron in stock, is estimated each week. If it grows above a two-week supply, the number of truck loads to be delivered in the next week is reduced. If it falls below one-week supply, the number of truck loads is increased. If the trend of consumption increases, or falls by fifteen per cent the delivery schedule is amended. The consumption of pig iron is a function of the number of orders received for castings.

8.3.3 *Bales of wool for a spinning mill (an implosive industry)*

All purchases of raw wool are made through a wool merchant. The mill management and the wool merchant cooperate to find a supply schedule which keeps the carding machines working, which gives an acceptable compromise between minimum cost and minimum stocks, and which meets the quality requirements set by the spinners. The flexibility needed to follow changes in market demand is obtained by regular meetings between the spinners and their wool merchant.

8.4 Ordering standards for assembled products

One of the main advantages of Period Batch Control (PBC) in explosive industries is that, because all orders are based on the same short-period production programme, the quantity of materials needed and the load on processing facilities each period are directly proportional to the number of products in the programme.

The factors which are directly proportional to the number of products in the production programme include:

(1) the number of parts *per product* to be made in each Group Technology (GT) group;
(2) the quantity of materials *per product* needed by each GT group;
(3) the net load in machine-hours *per product* on each machine tool, in each group;
(4) the number of bought parts *per product* to be delivered on call-off each period, by each supplier;
(5) the load in assembly man-hours *per product*, imposed by the assembly programme.

Standard values for all these factors (the ordering standards) can be stored in a computer data bank. Ordering, and the production of supporting data, will then only involve the simple multiplication of the ordering standards by the quantity of each product in the production programme each period. This produces, every period:

(1) list orders for each GT component group;
(2) a list of materials needed by each GT group;
(3) a load summary for each GT component processing group, in machine-hours per machine;
(4) a bought-parts list for each work station in each GT assembly group;
(5) a load summary for each GT assembly group, in man-hours per period.
(6) call-off notes for each supplier.

It should be noted that the standards are independent of the production control (PC) system. Any changes in the standards due to design changes, or changes in the processing method, can be introduced progressively during the week, without affecting the time required for ordering. The computer routines for ordering are therefore very simple and fast, and ordering can normally be completed on the Friday evening following the programme meeting.

8.5 The ordering schedule

The previous chapter looked at the schedules involved in flexible programming, for process, implosive, square, and explosive industries. If these flexible programmes are now used to regulate the ordering of parts or materials in

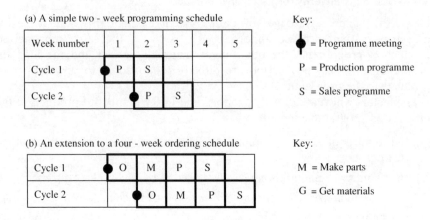

FIG. 8.2 From the programming schedule to the ordering schedule.

explosive industries, their PBC programming schedules will have to be amended to cover the additional ordering work. A number of examples from practice can be used to illustrate the changes that are needed.

8.5.1 Adding periods for component manufacture

Figure 8.2 shows a case where a simple two-period programming schedule has been extended to a four-week ordering schedule to allow time for obtaining materials and for component processing.

There is a programme meeting on Friday afternoon each week, and as soon as the production programme has been approved the computer produces:

(1) list orders for all groups covering parts manufacturing for the cycle;
(2) a list of materials needed for processing by each component processing group;
(3) a load summary for each group, showing the load in machine-hours on each machine and man-hours load for the operators;
(4) call-off notes for the suppliers.

These are all issued the same evening as the programme meeting; in the case of the call-off notes by computer link.

Materials are obtained in week 1, and made parts are processed in week 2. Bought parts for assembly, with a call-off lead time of two weeks, are received by the end of week 2. The products are assembled in week 3, and they are delivered to customers by the end of week 4.

8.5.2 Ordering schedules with more than four periods (weeks) per cycle

There are cases where incompatible processes must be carried out in succession at different processing stages. An example is illustrated in Fig. 8.3 for a

(a) First solution

Week number	1	2	3	4	5	6
Cycle 1	● O	F	M	A	S	
Cycle 2		● O	F	M	A	S

Key:
O = Order
F = Foundry
M = Machine
A = Assembly
S = Deliver
● = Programme meeting

(b) Second solution

Week number	1	2	3	4	5
Cycle 1	● F	M	A	S	
Cycle 2		● F	M	A	S

FIG. 8.3 The ordering schedule for a foundry, for a machine shop, and for assembly.

company making cast-iron castings in sand moulds in one period, ready for machining in the next period, and for assembly in the following period.

Any increase in the number of periods in the schedule increases throughput times and stocks, and this should be avoided where possible. In the case illustrated in Fig. 8.3(a), the final solution was to make the castings in week 1 (see Fig. 8.3(b)). A list order for castings is given to the foundry on Friday afternoons after the programme meeting. Simple castings without cores will be made on Monday of each week, allowing time to start core making for the more complex castings produced later in the week. It would be difficult in this case to combine a foundry based on sand moulds with precision machining, but there are several examples in practice where cleaner processes (such as die casting and hot pressing) are carried out in the same groups as machining.

Another example of a five-week schedule is given in Fig. 8.4 for a factory making taps and other bathroom and kitchen fittings. In this case, some parts were made from hot pressings and die castings. About half the parts produced, coming from five different component-processing groups, required chromium plating. One reduction in the original five-week schedule has already been achieved by including hot pressing and die casting in the same groups as machining (see Fig. 8.4(b)). A further reduction for the future is illustrated in Fig. 8.4(c). If machining is scheduled to machine the parts requiring electroplating early in the week, both machining and electroplating can completed in the same week, reducing the schedule to only three weeks.

8.5.3 Schedules for assemblies with long throughput times

Figure 7.3 showed a case where there were two successive assembly stages, both of which required a throughput time of more than one week. The product was a tractor-mounted digger. There were two assembly stages. In the first stage,

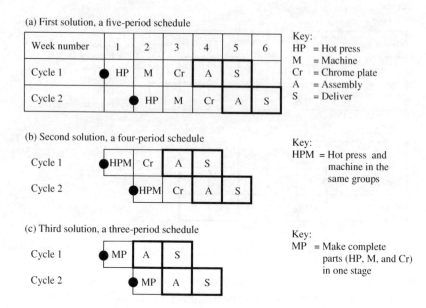

FIG. 8.4 Order scheduling for bathroom fittings.

the steel frame was welded, stress relieved, cleaned by shot-blasting, prime-painted and bored, and some holes were drilled in it. If, as shown in Fig. 8.5, one could reduce the throughput times for each assembly stage, from 3 days to 2 days, one could start the manufacture of the weekly requirement of frames

Cycle	Week number					
	1	2	3	4	5	6
1	Ac	P1 P2	S			
2			Ac	P1 P2	S	
3				Ac	P1 P2	S

Key:
P1 = Frame welding (solid line)
P2 = Main assembly (dashed line)
S = Delivery (starts Friday in week 2)

FIG. 8.5 The order schedule for a tractor digger. (See Fig. 7.4.)

on Monday morning in week 2. The first frame would be finished on Tuesday evening (week 2), and the last frame could be started on Friday afternoon in the same week and be finished by Tuesday evening in the following week (week 3).

The second or main assembly stage took the frame, plus a large number of made and bought components, and finished the assembly of the digger. If this assembly stage started on Wednesday morning in week 2, then on completion of the first frame the first digger for the cycle could be completed in two days, finishing on Thursday evening in week 2. The last digger would be started on Wednesday morning in week 3, and it would be finished by Thursday evening in week 3.

A large number of made components were produced in a machine shop. With very few exceptions, these were all required for the second stage of assembly. They were scheduled, for manufacture in week 2, ready for final assembly in week 2 and 3. Frame assembly and machining run in parallel in week 2 in different workshops. A few made losses were needed for frame welding, and some buffer stock of these losses was needed to ensure supply on time.

It might be said that success in this case could only be achieved by bending the PBC rules, because stages are not completed by the end of each period. It is true that this solution complicates progress control, but this was seen as a small price to pay for the lower stocks and shorter delivery time possible with the shorter schedule.

8.6 Summary

In the process and implosive industries, there are very few varieties of material used, and simple ordering methods based on the trend of actual usage and systematic monitoring can be used efficiently.

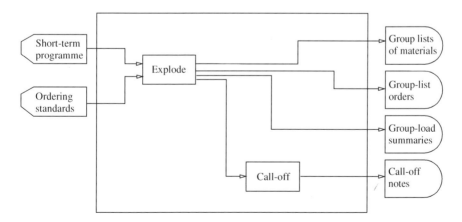

FIG. 8.6 A PBC ordering system Fo-explosive industries.

In the square industries, materials are normally provided by the customer, and only general processing materials have to be ordered by the manufacturer.

Ordering is only complicated in the case of the explosive industries, where parts have to be made or bought, and they must be ready in sets in time for assembly. The system diagram in this case is illustrated in Fig. 8.6. The general rule is that components should be delivered ready for assembly by the end of the period preceding assembly. The quantities to order, and the call-off from suppliers, are found by simple calculations on a computer, based on simple ordering standards. Ordering can usually be completed on Friday evening each week after the programme meeting.

It will be noted that PBC ordering is similar in principle to the Kanban system. The main difference is that PBC accumulates orders for one period before issue, while the Kanban system issues an immediate order for a replacement as each container load is issued to assembly. Kanban is mainly a method for mass production.

9

THE TIME CONSTRAINTS AND PBC

9.1 Introduction

Flexible or short-term programming has been used in the past to regulate assembly in explosive industries in cases where component processing and purchasing were treated as independent activities which were planned to maintain stocks. If parts are ordered to maintain stocks in a store using stock control (SC) ordering, or if parts are ordered by a multicycle system such as Materials-Requirement Planning (MRP), which is based on explosion from a long-term programme, and if these parts have to be sent to a store for kitting-out or for collection into assembly sets, then advantages can indeed be obtained, if, as a start, assembly is regulated by a series of short-term programmes.

If, however, there is an attempt to use this same flexible programming to regulate the production of piece-part products in implosive or square industries, or to provide the basis for explosion for ordering components in explosive industries—as is the case with PBC—then there are five major time constraints which have to be reduced.

1. There must be sufficient capacity, in machine-hours or man-hours, per period, to cover the load imposed by the period production programme.
2. It must be possible to complete one period's requirement of parts and assemblies in a throughput time of less than one period.
3. Any increase in the set-up frequency, due to the adoption of short periods, must be balanced by a reduction in set-up times, at least on bottleneck or other heavily loaded machines, or on machines which normally have long set-up times.
4. It must be possible to obtain purchase deliveries with a short call-off lead time of, for example, one period for materials and two periods for parts.
5. The sequence-constraint problem, imposed in some groups by a common sequence of operations, must be overcome. There are two main ways in which these problems can be tackled. The first starts by accepting and measuring the present values for the capacity; for the load, and for the throughput, set-up and purchase lead times, and then attempts to choose a period which can accommodate these constraints. Unfortunately, this approach does not generally find an efficient solution.

136 PROGRAMMING, ORDERING, AND DISPATCHING WITH PBC

The second way is to choose a period which will make it possible to achieve a required rate of stock turnover (say periods of one week), and then to study the present load and capacity, and the throughput, set-up, and purchase-lead-time constraints, and find where and how they must be changed, and by how much, in order to achieve the required rate of stock turnover. This chapter favours the second alternative, and shows how it can be achieved.

9.2 Capacity and load

In studying these time constraints, it is best to start by looking at the effect on the capacity and load in the production system of adopting Period Batch Control (PBC). The capacity and the load are both measured in terms of machine-hours and/or man-hours.

Capacity is the time available for productive work. For a machine, there is a maximum possible value of 168 machine-hours per week (7 days × 24 hours), but in practice, most machines will only be manned and made available for work during one shift, giving an available capacity of say, 40 machine-hours per week. For people, the capacity in man-hours is limited to the number of

Load Summary			Group name: C.I. turn	Group number: 7		Week number: 18 (1993)		
Machine				Capacity		Load		
Number	Plant number	Name	Quantity	Available	Net	Net	+	−
Tc 21	010119	Centre lathe	2	100	70	62	8	−
Tc 21	010120	Centre lathe						
MU 11	040504	Mill, universal	2	100	66	42	24	−
MU 11	040552	Mill, universal						
DC 2	051001	Drill coordinate	1	50	37	33	4	−
DC 4	050902	Drill coordinate	1	50	36	16	20	−
DT 61	051101	Tapper	1	50	45	23	22	−

FIG. 9.1 The load distribution on the machines in a group.

hours that they are willing to work. Forty hours, plus, say a maximum of 15 hours overtime per week, might be acceptable, and this might be established as normal in a particular factory.

The load is the time committed by orders, for work to be done at a work centre, using the available capacity. It is important to remember that the load will never be in balance with the capacity on all the machines in a factory or group. There will normally be one fully loaded bottle-neck machine, and the remaining machines will vary between heavily loaded and very lightly loaded, as illustrated in Fig. 9.1.

There are two main types of load. First, on any machine there is the *net load* per period, or the time (in a machine shop, for example) that will be spent cutting metal, plus the time for loading and unloading the machine. This is the sum of the operation times for each piece made on the machine, multiplied by the number of pieces made per period. Secondly there is the *gross load*, or the net load, plus the time required for setting up the tooling on the machine, plus an allowances for down time and for idle time when the machine is not producing useful work.

The net load is not difficult to predict, but the gross load imposed by an order is very difficult to predict, because the set-up time depends on the sequence in which parts are loaded on a machine. It can vary between zero (if the next job uses the same tools and the same set-up as the previous job) and several hours (if all the tools have to be changed or reset before the next job can start).

The most accurate method for capacity–load control is to measure the total nonproductive time per week on each machine (spent in setting up, or as down time or other idle time) using random-observation studies or machine monitoring. This value, or the average of recent values for each machine, is substracted from that machine's *available time* to find its *net capacity*.

The load on each machine in a period for comparison with this net capacity, is then the sum of the operation times for all the different items needed to meet the production programme, each multiplied by the quantity ordered per period.

The first limitation imposed by PBC is that the load imposed by the short-term production programme must never be greater than the available capacity for a period. If it is greater, the capacity must be increased or the load must be decreased, in order to increase the free capacity. Figure 9.2 shows some of the ways in which this can be done.

With production control (PC) systems other than PBC, this limitation does not apply because with some increase in stocks overload in one period can be carried forward to the next period. It is particularly important, therefore, with PBC, that the load imposed by orders in each period should be compared with the available capacity. It is a happy circumstance that with Group Technology (GT) and PBC it is very easy to make this comparison.

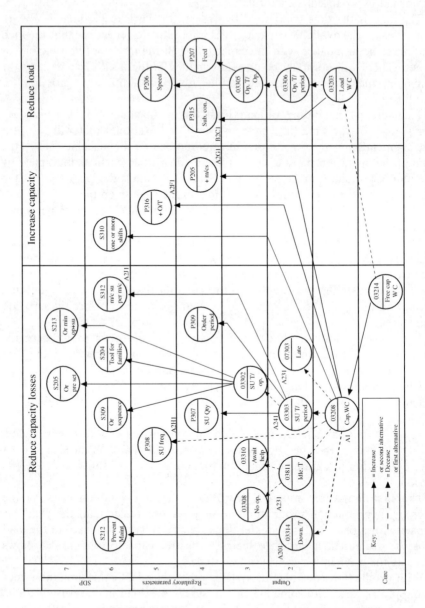

FIG. 9.2 Changes to increase the free capacity for a cure network. (See Fig. 3.2.)

9.3 Throughput time

The throughput time for a component, or for a batch of components, is the time it takes to pass through a specified segment of the materials-flow system (MFS). In a machine shop, for example, measurements can be made of the throughput time taken by a batch of parts passing through a department by subtracting the time when the batch of materials is issued to the workshop from the time when the last item passes final inspection and leaves the department. The manufacturing lead time, incidentally, can be defined as a forecast of future throughput times.

9.3.1 A throughput-time index

To compare throughput times, a fixed index per part is useful, this will not change with every minor variation in the load or in the sequence of loading. The most widely used throughput time index for parts used with PBC is the available time per period divided by the sum of the operation times for the part. It is not uncommon with process organization to find that the throughput time for a part is one hundred or more times greater than the sum of its operation times. With GT and PBC, the throughput time for one stage for a batch of parts must be less than one period. If the sum of the operation times for a part is one hour, and the available time for machining is one week of forty hours (a single shift), then the throughput-time index is 40. If the sum of the operation times is more than 40 hours, and the throughput-time index is less than one, this indicates that the working hours must be increased by working shifts, or that the throughput time must be reduced, before PBC can be used. It should be noted that this index is based on the sum of the operation times for a single part. If the period production programme calls for more than one part, the batch throughput time will be more than that for a single part.

In relation to throughput time, the critical parts are those with the largest values for the sum of their operation times; that is, those parts with the smallest throughput-time indices, and with the highest requirement quantities. These parts must be found and action must be taken to reduce their throughput times before the introduction of PBC, starting with those parts with a throughput-time index of, say, 10 or less.

9.3.2 Reducing throughput times

It is not difficult to reduce throughput times if the effort is made. The main methods are:

(1) GT;

(2) close-scheduling;

(3) process integration;

(4) shift working;

(5) reducing run quantities.

140 PROGRAMMING, ORDERING, AND DISPATCHING WITH PBC

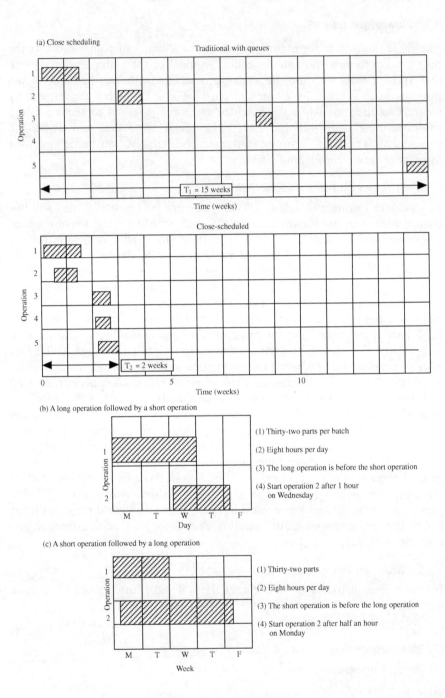

FIG. 9.3 Close-scheduling.

Group Technology Group Technology reduces throughput times by bringing the machines used to make parts close together and under one foreman. In practice, where this change to GT has been total, it has always at least halved the average throughput times previously experienced with process organization.

Close-scheduling Close-scheduling applies only to batch production. Assume that the normal throughput time for a part is the sum of the operation times per operation, multiplied by the number of parts required, with an allowance for queuing delays. Much shorter throughput times are possible if, as shown in Fig. 9.3, following operations are started on a batch before the preceding operations have been completed on all the parts. As shown in Fig. 9.3, the times at which following operations can start, if there is to be no lost time, depend on the ratio of successive operation times. To achieve these close-scheduling savings, the machines for successive operations must be laid out next to each other, or be connected by a conveyor, so that single parts can be transferred easily between operations.

Process integration Process integration, or reduction or the number of operations per part, is illustrated in Fig. 9.4 using *cells* for integration. In this case, three welding operations are followed by four operations on other machines, and this is followed by deburring. If the four machine tools are laid out next to each other to form a *cell* one worker can operate all four of them, giving major reductions in both the throughput time and the cost. A cell like that needed in Fig. 9.4 might be a special cell for one part, or it might be arranged to make a family of similar parts, or it might again be used for, say, seventy per cent of the parts made, leaving the individual machines free for other work on any remaining parts.

The operation route number (ORN) frequency chart, produced during line analysis (see Fig. 2.6), will quickly show if there are any possibilities for process integration. Process development can also integrate operations. For example, at the Cosworth engine factory, the development of methods for machining cylinder heads between centres has reduced the number of operations per part from 17 to 3.

Shift working Shift working does not reduce throughput times, but by making more working time available during a calendar period it makes it possible to include parts which require more machine time than is available in a single shift without changing the throughput time.

Reducing run quantities Reducing the run quantities reduces the throughput times, because the throughput time for a large run quantity is longer than that for a small run quantity. To reduce run quantities without reducing output, run frequencies must also be increased. This is the general effect of introducing PBC with short periods.

Route card		Part number: P44089	Name: Bracket		Group number: 8
Operation		Work centre		Op. T each	
Number	Name	Number	Name		
01	Weld frame	31982	Arc weld	20	
02	Weld bearing unit	31982	Arc weld	19	
03	Weld final	31993	Arc weld	15	
04	Mill feet	41/503	Mill vertical	17	
05	Drill 4 in feet	40/434	Pillar drill	9	
06	Drill 2	40/617	Radial drill	6	
07	Stud weld 3	31/998	Stud weld	10	
08	Deburr	71987	Bench	04	
				100	

Route card		Part number: P44089	Name: Bracket		Group number: 8
Operation		Work centre		Op. T each	
Number	Name	Number	Name		
1	Weld frame	31982	Arc weld	20	
2	Weld burring unit	31982	Arc weld	19	
3	Weld final	31993	Arc weld	15	
4	Machine	50/111	Machine cell	30	
5	Deburr	71987	Deburr	04	
				88	

FIG. 9.4 Process integration.

9.3.3 *Why throughput-time reduction is important for PBC*

Throughput-time reduction is important for PBC because PBC cannot be used unless all component throughput times per stage can be reduced to less than one period. This is partly a problem of how to reduce the throughput times for critical parts, using the methods described above, and partly of how to reduce delays due to interference between the cycles for different parts.

In the West, for over a century, the economic strategy for the development of manufacturing has been based on the reduction of operation times. This was accepted as being desirable, even if it increased throughput times. The Japanese developed the idea that reducing the throughput times also has a major effect in increasing the rate of return on investment and that it is possible to reduce throughput times without reducing operation times. Throughput-time reduction increases the return on investment by reducing stocks, which also indirectly reduces the costs of stock holding. The resultant increase in profit (due to cost reduction), and the reduction in investment (stocks), gives a geared increase in the rate of return on investment.

Figure 4.4 has already shown how changes in the throughput time affect the stock investment when the run frequency is fixed.

9.4 Set-up time

The third time constraint which is important for the successful introduction of PBC is that, at least in the case of heavily loaded machines, and for machines with naturally long set-up times, any increase in the set-up frequency imposed by the introduction of PBC, must be compensated by an equivalent reduction in the set-up time per set-up. If, for example, PBC is introduced with periods of one week, parts which were previously made every six weeks may now be made six times more frequently, and unless the time per set-up is reduced the additional set-up time may seriously erode the capacity.

Once again, however, it is not difficult to reduce set-up times if the effort is made. The following are the most effective ways of doing so:

(1) make-ready;
(2) sequencing;
(3) standardization;
(4) pre-setting;
(5) tooling development and machine modification;
(6) dedication of machines.

9.4.1 *Make-ready*

Very large reductions in setting time can be made by moving the tools, the drawings, and the operation sheet to a machine, ready for the next set-up, before the previous job has finished. It will be realized that advantage can only be taken of this possibility if the next two or three jobs on the machine are known in advance.

This tooling may be provided for a single part, or for a tooling family of parts. Figure 9.5 illustrates the case of a machine for which the foreman receives a matrix showing the parts in his group family which use the machine and may be ordered each period, divided into tooling families and showing the tools used to make them. Tools for each tooling family and other set-ups will be *made-ready* and issued to the machine before it finishes the previous job.

To take advantage of these make-ready savings, it may be necessary to store tools in the groups.

Tooling Families		Machine number: LC 721					Name: CNC lathe			Group number: 6		
	Tools	Part number										
Number	Name	080 59	001 57	019 01	001 89	011 22	011 16	001 71	000 04	004 01	300 06	656 09
–	Bar stop	√	√	√	√	√	√	√	√	√	√	√
232312	Part off	√	√	√	√	√	√	√	√	√	√	√
037702	Holder XX	√	√	√	√	√	√	√	√	√	√	√
037102	Holder XY	√	√	√	√	√	√	√	√	√	√	√
63.5	Collet						√	√	√	√	√	√
80.0	Collet	√	√	√	√	√						
120412	Tool tip square	√	√	√	√	√	√	√	√			√
	Tool tip round	√	√	√	√	√	√	√	√	√	√	√
2308.52	Tool tip copy			√	√	√	√	√	√			
0371025	Holder XZ	√	√	√	√	√	√	√	√		√	
3225.12	Holder facing	√	√	√	√	√	√	√	√	√		√
20550	Boring bar	√	√	√	√	√	√	√				
16	Centre drill									√		
32	Twist drill			√		√	√					
38	Twist drill				√							

Tooling family number 1 (columns 080 59, 001 57, 019 01, 001 89) → ⊕ Collet change

Tooling family number 2 (columns 011 22, 011 16, 001 71) → ⊕ Collet and drill change

Sequence of loading

FIG. 9.5 A tooling matrix for a group.

9.4.2 *Sequencing*

The setting-up time for a part on a machine depends on the nature of the previous job. If two very dissimilar parts are made one after the other, it may be necessary to change all the tooling on the machine, and the set-up time will be long. If two very similar parts, which use the same tooling, are made one after the other, it may be possible to make them both with the same set-up, and the set-up time for the second part will be zero.

Sequencing is the process of planning the loading sequence of parts on a machine in order to minimize the set-up time. In particular, sequencing looks for *tooling families* of parts, which can be made one after the other at the same set-up. If, for example, the average set-up time on a machine is 45 minutes, and it is found that ten parts can be made at the same set-up, the set-up time for these parts will be reduced to $4\frac{1}{2}$ minutes each.

Even if parts require different set-ups, it may still be possible to sequence the loading to reduce the set-up times. If, for example, on a lathe, changing the set-up between two parts only requires changing one or two tools, the set-up time will be much shorter than for a total change of tools. The tooling matrix in Fig. 9.5 shows the type of information needed for sequencing.

It is difficult to use sequencing with process organization and with multi-cycle ordering, because different parts which might be made with the same or similar set-ups are often made on different machines, and they are also seldom on order together at the same time. With GT, if planned with production flow analysis (PFA), similar parts which are suitable for sequencing are all likely to be made on the same machine, and with PBC all parts are ordered together at the beginning of each period so that sequencing is easy to arrange.

9.4.3 *Standardization*

Standardization, particularly of materials and of design features, can lead to significant savings in the set-up time. The standardization of a limited range of steel-bar diameters for machining on bar lathes, will reduce the set-up times by reducing the number of bar feed and collet changes needed. This can be done without changing the product design, if some additional waste is allowed when parts are machined from bars with a slightly larger diameter than the nominal value for the part.

Figure 9.6 shows how standardization of the tools for sheet metal bending reduces the setting-up time on a brake press. Figure 9.6(a) shows how important savings were made by sequencing, and Fig. 9.6(b) shows how reducing the number of press tools with different radii, and reducing the number of V-notch dies increased the size of the tooling families, giving a further reduction in the setting-up time.

Figure 9.7 shows a case where a cylindrical grinder has to grind a cylindrical form which blends into a radius. If this radius is standardized for a range of parts, they can all be ground with the same radius on the grinding wheel, thus reducing the set-up time.

Standardization of design features can also be seen as the Standardization of the special cutting tools used to produce them. Standardization not only reduces set-up times, it also reduces the investment in cutting tools.

146 PROGRAMMING, ORDERING, AND DISPATCHING WITH PBC

(a) Before standardization

Bottom / Radius (mm)	50	80	125	150	200	Form
12	(11)					
16	9 ✓					
20		(14)				
25		23 ✓				
30		3 ✓				
32			(6)			
40			4 ✓			
45			4 ✓			
50			3 ✓			
60				1 ✓		
75					(5)	
83					1 ✓	
100					1 ✓	
148 B				2 ✓		
Form						(1)

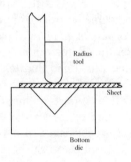

88 parts

60 min set-up time, or
45 min, change one tool
◯ = 60 min set-up

Random loading SU = 990 min
Sequenced SU = 765 min

(b) After standardization (to reduce the number of tools)

Bottom / Radius (mm)	Small 50,80	Medium 125	Large 150,200	Form
16	(20)			
25	40 ✓			
40		(17)		
75			(8)	
148 B		2 ✓		
Form				(1)

88 parts

Sequenced SU = 330 min
after standardization

FIG. 9.6 Standardized dies for a brake press (SU is a standard unit).

9.4.4 *Presetting*

A more expensive way of reducing set-up times is to preset the tools, using special tool holders and fixtures, so that any tool can be replaced on a machine in a matter of seconds; once replaced, the new tool will be in the correct

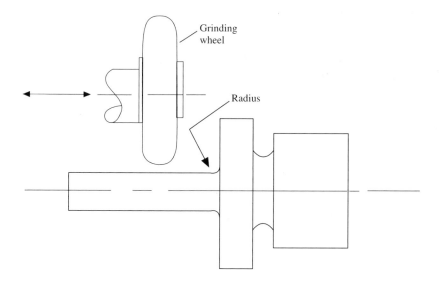

FIG. 9.7 Standard radii for shaft grinding.

position to machine the materials to the exact dimensions and tolerances required by the product design.

It should be remembered that thirty or forty years ago setting-up a tool involved positioning the tool in roughly the position needed to cut materials to the right size. In clamping, the setter then attempted to fix the tool onto the machine in the correct position. Setting-up was very much a process of trial and error and adjustment.

The Japanese writer Shigeo Shingo, in his book *Single minute exchange of dies* (SMED) sees this approach of presetting in a different way. He sees the tasks of setting-up, as divided between *internal tasks*—performed on the machine while it is stopped—and *external tasks*, which can be done off the machine while it is working on some other part. These external tasks include not only collecting together the tools and data needed for the next set-up (make-ready), but also the presetting of tools while the machine is doing useful work on some other part. Shigeo Shingo sees set-up-time reduction as largely a problem of changing internal tasks into external tasks.

It will be noted that the set-up-time reductions obtained in this way reduce the time when a machine is not producing useful work. It does not necessarily reduce the cost of setting, if the cost of presetting is taken into account. Once again, to obtain flexibility and accountability, presetting and tool storage are best allocated to the groups.

The reduction in set-up times has some absolute value, in that it can reduce the expenditure on setting-up, and that it increases capacity. Research has indicated, however, that the main value of set-up-time reduction is that it

148 PROGRAMMING, ORDERING, AND DISPATCHING WITH PBC

makes it economical to manufacture with low run quantities and a very low investment in stocks and work in progress (WIP).

9.4.5 *Tooling development and machine modification*

Another way of reducing set-up times is by tooling development. Figure 9.8 shows how the late Signor Patrigani (an Italian engineer, working as a consultant in the 1950s, for the French company Societe Alstham, at Lecourbe near Paris) reduced the set-up times on 90 ton power presses from 45 minutes to 15 seconds.

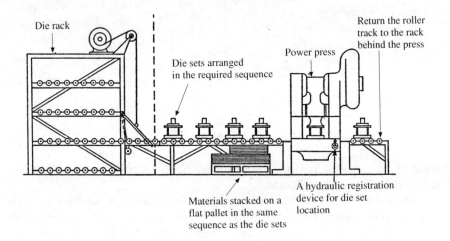

(a) Set-up time: 5 seconds

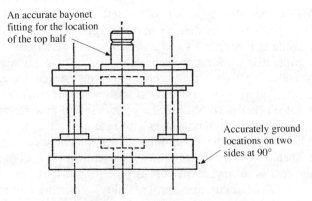

(b) The modification of a die set by the method of Signor Patrignani

FIG. 9.8 Reducing the set-up time on a power press.

The die sets were ground with location surfaces at 90°, accurately positioned in relation to the bayonet connector on the top of the die set. An accurately ground register and retractable location studs at 90° were provided on the press to locate the die sets. The materials were stacked on a flat palette by the machine in the sequence of usage, and the die sets were arranged on a roller track leading to the press, also in the sequence of usage.

When the stack of material for a part had been used, a hydraulic clamping device was actuated to release the die set. This die set was rolled to one side, and the next die set was rolled onto the machine and located by pushing it against the stops, into the correct position. It was then automatically clamped in that position. Work was then started on the next batch of material.

9.4.6 *Dedicated machines*

Another method of reducing set-up time is to leave a machine permanently set-up for one particular operation which has a long set-up time. It sometimes happens that when a machine tool is replaced by a more modern machine, the old machine has little more than scrap value. It may be economical to retain the old machine, leaving it set up permanently for some operation on a part which has a long set-up time. In a recent study in a general machine shop, make-ready, sequencing, and dedicated machines—which are the low-cost methods of set-up-time reduction—achieved eighty five per cent of the set-up-time reduction needed to introduce GT and PBC with one-week periods. Presetting gave a further ten per cent reduction leaving five per cent requiring new tools or machine modification, or product design changes.

9.5 The sequence-constraint problem

One other problem caused by the use of PBC, is the sequence-constraint problem illustrated in Fig. 9.9(a). If all the parts made in a group use the same machines in the same sequence, and if these machines are evenly loaded, there may be capacity losses on some machines due to lack of work at the beginning and at the end of the ordering period. In most groups, this problem does not arise because there is spare capacity on most of the machines, and because there is sufficient variety in the routing sequences so that more than one machine does first operations, and it is possible to find work for all machine operators early in each cycle. When this problem does arise, the losses can be reduced by close-scheduling when the following operations start before the previous operations have been completed on all the parts in a batch (see Fig. 9.9(b)). Note that the larger is the number of parts with the same sequence of operations, the easier it is to eliminate the sequence-constraint losses. These losses can also be reduced by reducing the number of operations per part, (process integration, see Figure 9.9(c)). The process planning objective of *single-shot machining*, eliminates the sequence-constraint problem completely.

150   PROGRAMMING, ORDERING, AND DISPATCHING WITH PBC

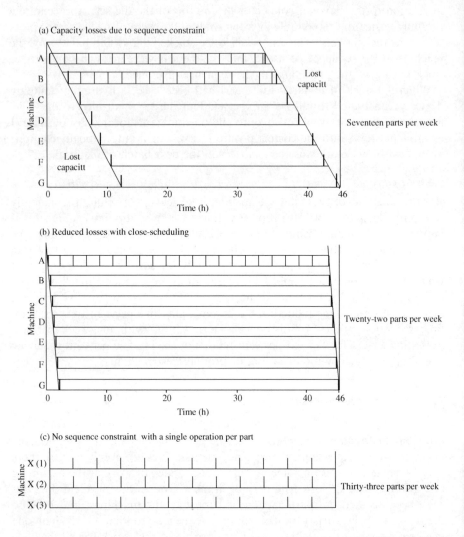

FIG. 9.9 The sequence-constraint problem. Note, the case where a common sequence gives maximum capacity loss is also the case where close-scheduling is easiest to use.

An example of the sequence-constraint problem in assembly has already been illustrated in Fig. 7.3, for a tractor-digger assembly. In this case, the sequence of frame welding stress relieving shot-blasting, and painting and machining make throughput-time reduction difficult for the manufacture of digger frames, the throughput time for the main assembly was also difficult to reduce. In this case, some overlap between periods was allowed and planned.

9.6 Purchase lead times

The final, and often the limiting, time constraint, which determines whether PBC can be used, is the purchase lead time. The method of purchasing used with Period Batch Control is known as the call-off method. With this method, suppliers are given long-term supply contracts covering the requirements for six months or more into the future. They are given a forecast, found by explosion from the annual production programme, of probable future requirements, but they are told that actual deliveries can only be made against periodic (say weekly) call-off instructions.

The lead time for obtaining deliveries against a new order may be as much as six months, allowing for process planning and the acquisition of tooling and special materials. The lead time for call-off against a long-term contract can be very much shorter. The supplier should need very little stock of an item to be able to deliver to call-off with a one-week delivery lead time (see Fig. 8.1).

9.7 Summary

Before PBC can be used, it is necessary to reduce five time constraints. These are:

1. The load per period must not be more than the capacity of each group.
2. The throughput time per part, per stage, must be less than one period.
3. The set-up time must be reduced to compensate for any increase in the set-up frequency on critical machines, due to the adoption of a single short period.
4. In some groups it may be necessary to eliminate capacity losses due to sequence constraints.
5. The call-off delivery lead time must normally be not less than one week, although shorter lead times may be economical for large quantities.

Experience, notably in Japan, has shown that it is generally possible to eliminate these constraints. We in the West have preferred to measure the constraints, and have then attempted to design PC systems which will accept the present constraint values. The success of Japan in manufacturing indicates that we need to think again.

10
DISPATCHING WITH PBC

10.1 Introduction

Dispatching is the third level of the function of production control (PC). It follows after *programming* and *ordering*, and covers the planning, direction, and control of the work done in company workshops to complete shop orders economically, to the required standards of quality, and by the specified due dates. It deals in other words, with the work done at the level of supervisory management, or of the foreman.

Typical tasks at the dispatching level in, as an example, an engineering machine shop with Group Technology, (GT), include:

(1) setting up;

(2) tooling storage and maintenance;

(3) inspection;

(4) materials handling;

(5) plant maintenance—a part of;

(6) swarf removal;

(7) good housekeeping;

(8) records;

(9) operation scheduling.

Operation scheduling, or planning the starting and finishing times for machine and manual operations, is the key process in dispatching. It will be examined last, however—in the next chapter—because the efficiency of operation scheduling depends on the other dispatching tasks; they need to be understood before work can be scheduled efficiently.

It is highly desirable with GT that the authority to manage dispatching should be delegated to the foremen, or group leaders, and that all related tasks which directly contribute to group efficiency should be done in the groups. If this is achieved and if the groups complete all the parts they make without back-flow (between groups at different stages) or cross-flow (between groups at the same stage), then the group leaders can fairly be held responsible for product quality, for those elements of cost which they can control, and for completion of the work by the due dates set in the programmes and shop orders issued by PC.

Attempts to centralize services, such as setting up, tool storage, inspection, materials handling, and floor sweeping, reduce the ability of the foreman to control the destiny of his group and tend to increase bureaucratic overhead costs for administration and control by more than can possibly be saved in the direct cost of the service by specialization.

10.2 Setting

One of the principal dispatching tasks is to regulate the setting up of machines. When an operation on a part, or on a tooling family of parts with similar set-ups, has been completed on a machine, some or all of the tools on that machine will usually have to be changed before starting work on the next operation.

Because stock reduction is a major objective of GT and Period Batch Control (PBC), there is a tendency to reduce the run quantities by choosing short periods for PBC. This will induce an increase in the run frequency, which in turn may also induce some increase in the set-up frequency. In this case, to avoid a loss in capacity, it is necessary to reduce the set-up times per set-up. The cheapest and most efficient methods for reducing set-up times depend on the dispatching efficiency. They can be described as:

(1) make-ready;

(2) sequencing;

(3) presetting.

10.2.1 *Make-ready*

The objective of make-ready is that, before any processing run is finished on a machine, the tools, drawings, and any information needed for the following set-up should have been delivered to the machine so that work can start immediately on this next set-up. With traditional methods of setting up, collecting the tools for a new set up was done after the previous job was finished. The time needed for this work was often fifty per cent or more of total set-up time.

To do this make-ready process efficiently with GT, it is desirable that the tools needed should be stored in the groups, and that there should be some worker or workers employed in each group who have been trained in the make-ready process.

It will be obvious that this work is operation schedule dependent. Make-ready for setting up cannot begin until it is known which are the next jobs on the machine. What is more, if presetting is used on the machine, the next job must also be known long enough before the previous job ends to allow time for the presetting.

10.2.2 *Sequencing*

The set-up time on a machine depends to a large extent on the nature of the previous job, and on how many tools have to be changed. Sequencing is the art of planning the sequence of loading work on the machines in groups to minimize set-up time.

The kits of tools for future set-ups may be for one single part or assembly, or they may cover a tooling family of different parts all made at the same set-up, or, finally, they may provide a kit of tools for a family of parts made with closely related but different subsets of this kit.

Figure 10.1 shows an example of this third case. Fourteen different parts are made on a lathe with ten different tooling positions. Two of the tools are common to all parts (1 and 11), but eleven of them may have to be changed when there is a part change. The parts are listed in Fig. 10.1 in the sequence which will give minimum set-up time. Some changes in this sequence can be made inside the tooling families when necessary, with little consequent increase in the total setting time. This makes the system flexible in the face of unforeseen production mishaps.

Sequencing can be seen as a major element in efficient operation scheduling, it can be planned before operation scheduling starts, using the tooling analysis (TA) subtechnique of production-flow analysis (PFA).

10.2.3 *Presetting*

Presetting requires that tool holders on machines should be made with a high degree of precision, so that if tools are accurately preset in the tool holders fixing the holders on the machine will accurately position the tools for the required operation.

As an example, a drill used on a capstan lathe to drill a hole to the required depth will be held in a special holder. The position of the drill in its holder, which determines the depth of the hole in the drilled part, is preset and fixed before the set-up starts. The simple operation of attaching the drill in its holder to the machine quickly and accurately positions the drill, ready for machining.

Presetting will usually require some investment in new tools and holders, and in some cases it may also require some modification of the machine tools. It is still, however, a relatively low-cost method for reducing the set-up time on a machine, that is, for reducing the *internal* set-up time. It may, however, have little effect on the total setting-up cost if it only transfers internal set-up time to external set-up time.

10.2.4 *Other methods for reducing the set-up time*

There are other methods for reducing the set-up time which are independent of dispatching and operation-scheduling skills. These include: standardization of materials and of design features, the use of dedicated machines for operations with long set-up times, changes in processing methods and tooling, and modifications to machine tools. These methods have already been described in Chapter 9. Since they are not affected by dispatching efficiency, they need not be considered again here.

	Sequence planning			Machine Number TC107												Group number: 3 Name: CNC lathe	
				Tools													
TF	Sequence number	Part Number	Part Name	Operation number	1 Centre tool	2 Collet (a)	3 Collet (b)	4 Collet (c)	5 Turn OD (a)	6 Turn OD (b)	7 Turn OD (c)	8 Bore (a)	9 Bore (b)	10 Drill (a)	11 Drill (b)	12 Tap	13 Turn thread
1	1	A7602	Stud	2	✓	✓			✓	✓		✓		✓	✓	✓	✓
	2	D176	Bolt	1	✓	✓			✓	✓		✓		(−)	✓	✓	✓
	3	G1001	Stud	1	✓	✓			✓	✓		✓			✓	✓	✓
2	4	A7698	Shaft	2	✓	(−)	✓		✓	✓		✓	✓		✓	✓	(−)
	5	G3261	Shaft	2	✓		✓		✓	✓		✓	✓		✓	✓	
	6	D188	Stud	2	✓		✓		✓	✓		✓	✓		✓	✓	
3	7	J1023	Shaft	2	✓		✓		✓	(−)	✓	✓	✓		✓	✓	
	8	J1097	Shaft	2	✓		✓		✓		✓	✓	✓		✓	✓	
	9	A8210	Bolt	1	✓				✓		✓	(−)	✓		✓	(−)	
4	10	G5059	Stud	1	✓			✓	✓		✓		✓		✓		
	11	D9096	Spacer	1	✓			✓	(−)		✓		✓		✓		
	12	A9913	Spacer	1	✓			✓			✓		✓		✓		
	13	G9009	Spacer	1	✓			✓					✓		✓		
	14	J3332	Stud	1	✓								(−)		✓		

Key: (−) = delete tool, ✓ = add tool, TF = tooling family, OD = outside diameter

FIG. 10.1 Sequencing.

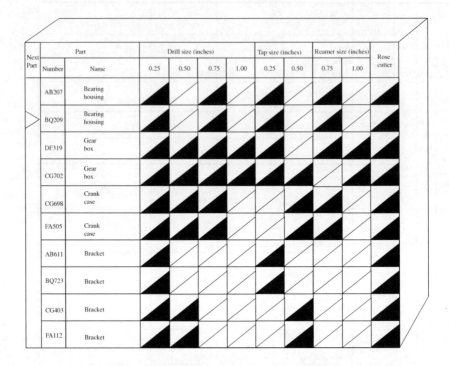

FIG. 10.2 A tool-store rack for a radial drill.

10.3 Tooling storage and maintenance

With GT, the tooling for setting-up machines should normally be stored in the groups, and this is particularly desirable with PBC, due to the high order frequency with short-period ordering.

If a machine uses very few tools, and/or there are very few parts which use the machine, it will normally be best to store the tools by the machine. Figure 10.2 shows the tool-storage rack for a radial drill in a group which uses nine tools to make ten different parts. When a new part is to be machined, the pointer at the side of the rack is moved to indicate which part is next. Tools are rearranged on the rack to fill all the required-tooling spots for the part, and a new drill jig is placed on the machine bed. Arrangements are needed for the regular inspection of tools and the replacement of worn tooling, but this method of storage makes it possible to change set-ups in a very short time.

With more complex machines, using many different tools, and/or machines which make many different parts, it may be more convenient to store the tooling on racks near a central *presetting point* in the group (see Fig. 10.3). The tooling records for the machines again take the form of a matrix, like that already illustrated in Fig. 10.1. Sets of tools are preset in the group and

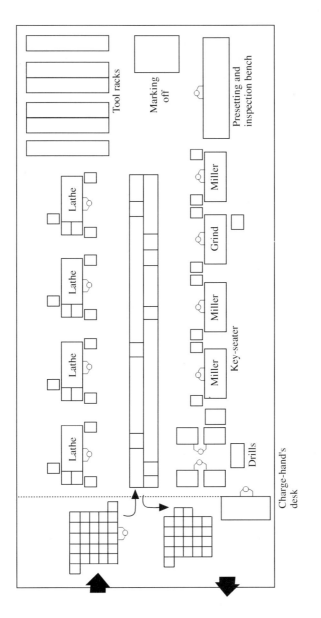

FIG. 10.3 Tool stores in a group.

delivered to the machine when needed. It is obvious, in this case, that before setting up it is necessary to know in advance what the next job is, or, if

Machine			Group number	Tool storage	
Item	Number	Name		By machine	Group tool store
1	2073	Face and centre	1	✓	
2	2156	CNC chuck lathe		Tool bank on machine	
3	8100	Cylindrical grind		✓	
4	4653	Universal mill			✓
5	3302	Two-spindle drill		✓	
6	6616	Jig borer		✓	
7	–	Ultrasonic cleaner		–	–
8	2710	CNC bar lathe	2	Tool bank on machine	
9	2718	CNC bar lathe		Tool bank on machine	
10	4666	Universal mill			✓
11	3391	Three-spindle drill		✓	
12	–	Ultrasonic cleaner		–	–
13	2143	CNC chuck lathe	3	Tool bank on machine	
14	2434	Centre lathe			✓
15	2810	Copy turn lathe			✓
16	4671	Universal mill			✓
17	3397	Three-spindle drill		✓	
18	3715	Tapper		✓	
19	–	Ultrasonic cleaner		–	–
20	2149	CNC lathe	4		✓
21	2456	Centre lathe			✓
22	4692	Universal mill			✓
23	3816	Radial drill		✓	
24	3101	Pillar drill 15 p		✓	
25	–	Ultrasonic cleaner		–	–

FIG. 10.4 Tool storage location in a machine shop.

operation times are short, what the next two or three jobs are which are scheduled for operations.

For many machine types, future storage will be in *tooling banks* on the machine and automatic tool changing will be used with instructions provided by a computer program. Figure 10.4 shows the twenty-five machine types and other work centres in a machine shop, and it also the shows location chosen in each case for tool storage.

The regrinding of tools is sometimes centralized in the tool room on a tool-exchange basis. Where the tools are used in several different groups (for example, drills), this is often a reasonable solution. There are cases, however, where the regrinding of tools is fundamental to the quality of the work produced, and the process must be included in the group concerned, if the foreman is to be made responsible for the quality of the work. As an example, there are several cases in practice where the grinding of gear hobs has been included in GT gear groups for this reason.

10.4 Inspection

Before the Second World War, inspection was seen as a service which should be independent from processing. More recently, the idea has grown that product quality depends on the way in which the process workers do their work. It is better, therefore, to educate workers to understand the importance of quality, and to teach them how to inspect their own work.

In practice, if inspection is very complex, it may be necessary to provide an inspection station manned by a skilled inspector inside some groups. Most of the work, however, will be inspected by the workers themselves. They should be trained in inspection, and they should be provided with any gauges or other tools needed to do the work. To maintain high standards of quality, most companies will need a *quality-control department* to recheck random samples of the work produced by the groups, in order to control whether they are inspecting their work effectively.

Group Technology, coupled with good plant lay-out (which lays out, as *cells*, sets of machines inside groups which are normally used in the same sequence), tends to reduce the number of inspections required. As a simple example, consider a drilling machine and a tapping machine. With traditional process organization, drilling of all parts in a batch will be finished before tapping starts. It will probably be necessary to inspect the work of drilling to ensure that the right size and depth of hole has been drilled. If, on the other hand, the drilling machine and the tapper form a cell inside a group, manned by a single worker who taps each piece while the next part is being drilled, it will be immediately apparent with the first piece, if an incorrect size of drill is being used or if the hole is not deep enough. A special inspection operation at this point in the process is unnecessary.

Group Technology coupled with PBC normally involves processing in relatively small batches. In this case, inspection of the first and last pieces in each batch is usually sufficient to maintain quality standards. Statistical quality-control methods are desirable mainly for long runs, or for continuous processes.

10.5 Materials handling

With GT, a common convention is that departmental transport delivers materials to groups and collects the finished work for onward transmission. All materials handling inside a group is the responsibility of that group. They should be provided with any handling equipment necessary for lifting and moving batches of parts.

As far as possible, standard containers should be used for handling, and space should be provided for materials by each machine, or set of machines (cell), installed in the group. One way of organizing handling is to provide a signal light on each machine, which is switched on by the operator, to indicate when containers are full and when they require movement inside the group. Seeing the light, the labourer reads the *follow card* to discover which machine does the next operation, and then switches off the light and moves the container to the next machine.

Figure 10.5 shows an arrangement for storing containers by the machine, in which they are stacked to save space. With some scheduling systems, the parts are machined on each machine in the sequence of their arrival. With other systems, this sequence may be varied. This problem is considered again in the next chapter.

10.6 Plant maintenance

With GT, the foreman of each group is normally responsible for the maintenance of the machines in that group. The foreman will usually be required to regulate daily and monthly machine inspections for preventive maintenance, using workers who have been trained to do the work and who are employed in the group. The longer term, say annual inspections, will be done by skilled machine-tool inspectors who will visit the groups when requested. Machine overhaul and major repairs will usually be centralized as a service available to the groups when they need it.

10.7 Swarf removal

In machine shops, the waste material produced by machining is known as *swarf*. It has some value as a raw material which can be melted down to produce new materials, also, when it is removed from machine tools, it tends to carry with it large quantities of cutting fluids or oils. These liquids can be recovered for reuse by filtration, or by centrifuging.

The coolant-recovery processes are normally organized as departmental processes which serve the groups. Since they are normally dirty processes required by several different groups, this is a reasonable approach. However, in cases where the service only relates to one group, which reuses the cutting oil after cleaning, it is better to install the service as a part of the affected group.

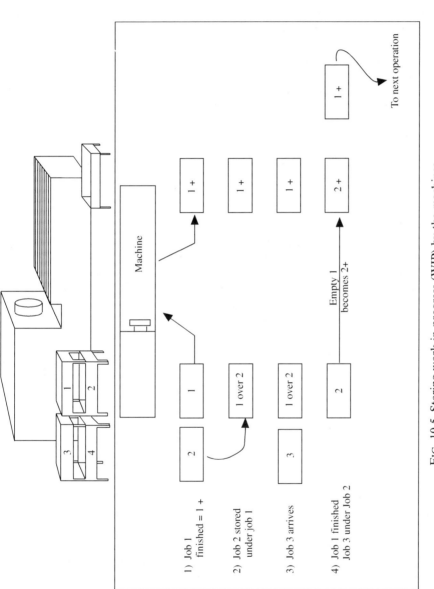

FIG. 10.5 Storing work in progress (WIP) by the machines.

10.8 Good housekeeping

The sweeping of floors, the cleaning of machines and other housekeeping jobs are again normally delegated to the groups. The cleanliness and tidyness of a group is a good indicator of its efficiency. Some companies make regular inspections of their groups, assess them for general cleanliness, and they publish these assessments to encourage high standards.

A few companies have employed subcontractors for cleaning who provide housekeeping as a service. It is doubtful if this approach is ever less costly than doing the work in-house, and the lack of responsibility for cleanliness in the groups tends to generate untidy habits.

10.9 Records

With process organization and multicycle ordering, detailed records are needed to show the state achieved by orders. In other words, it is necessary to know how many parts have been completed at each operation, before the present state of an order can be known.

With GT and PBC this information is unnecessary. All orders cover the requirement for one short period, and in each period all the orders have the same order date and due date. Under these conditions the *present state of any order* is concerned with one GT group, and with one particular (say, weekly) time period. There is no advantage to be gained by recording the daily progress of such orders. The foreman must be able to see and understand this daily succession of states, but he does not need detailed records to do so.

The only records needed in a GT and PBC situation, are periodic group-performance records, covering for example:

(1) parts not completed by the due dates (that is, by the end of each period);

(2) the number of quality rejects expressed as (say) *rejected parts per million*;

(3) absenteeism in man-hours lost per period;

(4) machine down time, in machine-hours, lost per period.

These methods are considered in detail in a later chapter (Chapter 17).

10.10 Summary

Dispatching is the third level of the function of production control. It deals with regulation of the work done in manufacturing workshops to complete parts and products. It is concerned with such tasks as: setting up, tool storage and maintenance, inspection, materials handling, plant maintenance, swarf removal, good housekeeping, and production records. The key process in dispatching is operation scheduling. This will be studied in detail in the next chapter.

With GT, the responsibility for dispatching is normally delegated to the groups. Period Batch Control tends to simplify dispatching, but at the same time it increases the need for reliability.

11

OPERATION SCHEDULING WITH GT AND PBC

11.1 Introduction

Production control (PC) is the scheduling function in production management. Production schedules are planned progressively in PC at three levels: first, by *programming*, which schedules the output of finished products; secondly, by *ordering*, which schedules the output of made parts from workshops, and the input of bought parts and materials from suppliers needed to complete products; and, thirdly, by *operation scheduling* in dispatching, which schedules the completion of work operations, at work centres in workshops, needed to complete orders by due date.

Generations of factory managers and of academics have sought a solution to the problems of operation scheduling. The many papers on the subject, which are still being published in production journals, illustrate the fact that a satisfactory solution has not yet been found.

This chapter submits that: operation schedules are extremely difficult to plan and control in factories with process organization and multicycle ordering. Operation scheduling is much simpler in factories with Group Technology (GT) and Period Batch Control (PBC), because these methods greatly simplify the materials-flow system (MFS). The method of operation scheduling known as *launch-sequence scheduling* (LSS) can be used in this case. Because it is simple, this method of scheduling can be delegated efficiently to the foremen of GT groups, as a part of their duties.

11.2 The scheduling problem

There is a great difference between operation scheduling with process organization and multicycle ordering, and launch-sequence scheduling (LSS), (which is based on GT and single-cycle ordering (PBC)). In the first case, each part ordered has a different order date and due date. In the second case—with GT and PBC—all parts ordered in each cycle have the same order date and due date. Orders are found by explosion from a series of short-term production programmes, often, currently, for periods of one week. Such shop orders, or group-list orders are normally composed mainly of parts for products for which there are existing sales orders. Unlike the orders found by stock control and Materials-Requirement Planning (MRP), such orders seldom require revision after issue.

It will be remembered that GT itself provides a significant measure of simplification, compared with process organization. Instead of one large complex schedule for the whole department, separate, smaller and simpler schedules are now needed, one for each group.

Again, workshops divided into GT groups are normally more reliable than those with process organization, because the groups complete parts under one foreman. Groups generally suffer fewer quality rejects, and fewer late completions for this reason.

Again, in machine shops with process organization, operations on common machine types (Type C)—such as lathes, mills, and drills in machine shops—may be allocated to any one of a number of different machines of the same general type. With GT, these common machines are divided between the groups. Groups tend to bring together parts with a similar shape or function, because they use the same S or I class machines, or because particular methods and machines have become established for particular types of part in a company. The reallocation-of-operations-to-machines step in group analysis (GA) brings parts with a similar shape or function together on the same C class machines. This increases the size of tooling families and helps to reduce setting-up times.

The group foreman receives with his periodic *list order* a *load summary* showing the load and capacity in machine-hours and/or man-hours for his group. With GT and PBC, it is simple to produce such summaries. The foreman knows before work is started if there is sufficient capacity to complete the orders, or if he needs to work overtime.

Foremen will also be given any other data that is needed if they are to schedule the work themselves. They will not be given a complete schedule for the period, for all the machines in their groups, partly because such schedules are expensive to produce, partly because they tend to be inflexible, and partly because, with GT and PBC, there are other much simpler ways to regulate the flow.

The main question to be addressed by foremen in order to regulate the work in their group, is: 'What is the sequence in which parts should be started on those machines which do first operations?'. Because each group makes its own fixed *family* of parts, this loading sequence tends to stay much the same, period after period, even though not all the parts in a family may be produced every period.

If there is sufficient capacity to cover the load, it will usually be possible to find a sequence for loading the machines for first operations, which, coupled with strict queuing discipline for following operations, will cause all parts on order to be completed by the end of that period. This method of operation scheduling is called launch-sequence scheduling (LSS).

11.3 Factors in planning the launch sequence

The primary objective of LSS is to complete all the parts ordered on each group by the end of the period. A number of secondary objectives can be recognized which must be achieved if the primary objective is to be achieved. To complete all parts by the end of each period, the work must be scheduled so that:

1. All workers are gainfully employed, quickly, at the start of each period.
2. All parts ordered are completed in one period.
3. Operations are scheduled for the minimum set-up time.
4. Bottleneck machines are kept fully loaded.
5. Parts for any processes with batch-processing operations must be accumulated in good time, ready for batch processing.
6. The schedule may have to be planned to load related services economically.

Each of these secondary objectives may tend to generate a different operation schedule. The aim is to find a compromise schedule which completes all parts by the end of the period, in the most economical manner.

11.3.1 *Finding work for group workers*

In any factory, only the bottleneck machine can be fully loaded. If the other machines are fully loaded to the limit of their capacity, a part of their output will be excess stock. Most groups have more machines than workers. There will inevitably be some *machine idle time*, but the *operator idle time* should be much less. It is essential to schedule the work so that all the workers are fully employed from the start of the period.

If there are more machines in a group which do first operations than there are operators, there is no problem. If, however, there are only a few machines which do first operations, some operators can be employed at the beginning of the period on presetting or setting-up machines, but the schedule must quickly find work on other machines to keep them all in work.

Figure 11.1 shows a machining group in which only four machines do first operations (CQ, CN, CP, and CT). These machines are linked to others in a line flow cell (cell 1), where close-scheduling can be used. Due to large differences in operations times, one operator can often work two or more machines. Nevertheless, in this group, work on machines is found for all operators within ten minutes of the start of the periods. There is a total of ten machine operators, and, allowing for the needs for machine setting and for inspection, there is little or no operator idle time in this group.

11.3.2 *Completing all the parts in each list order in one period*

With PBC, all parts ordered in each period must be completed by the end of that period. This is not a difficult problem for simple parts, with few operations and short operation times. The problem is only difficult for a limited number of critical parts, with many operations and long operation times. Figure 11.2 uses Pareto analysis to show the distribution of throughput times in a particular group. It is desirable that work on the critical parts—eight per cent

166 PROGRAMMING, ORDERING, AND DISPATCHING WITH PBC

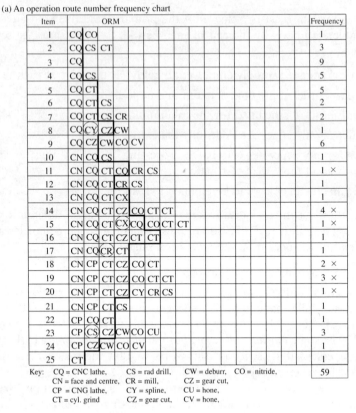

(a) An operation route number frequency chart

(b) The division of groups into cells

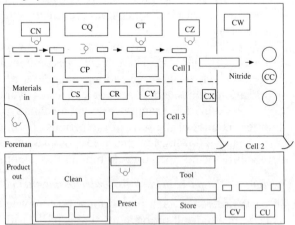

FIG. 11.1 A machining group.

of the total—should be started early in each period, to give plenty of time for completion.

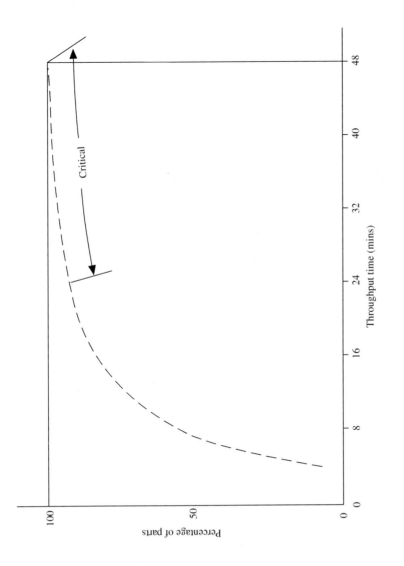

FIG. 11.2 The distribution of throughput times in a group.

After GT, the best way to reduce throughput time (as described in Chapter 9) is to close-schedule operations so that following operations are started as soon as possible after the first part in a batch has been completed on the previous operation. This requires that the machines in a group, which are normally used to perform a sequence of operations, should be laid out next to each other in *cells* to facilitate unit material transfer (see cell 1 in Fig. 11.1).

Close-scheduled cells are, in effect, lines. Station load balancing is, however, usually impossible in such cases. The lines can only be balanced by moving labour from one work station to another, in order to keep the most heavily loaded machines fully occupied and to minimize the accumulation of work in progress between machines.

11.3.3 *Scheduling for the minimum set-up time*

Another factor which is closely related to scheduling is the set-up time. Each group produces a fixed family of parts, and each machine in the group is used to perform a fixed list of operations on some, or all, of these parts.

Tooling analysis (TA) can be used to find sets of parts known as *tooling families*, with operations which can all be done on a machine using the same set of tools. On many machine types, it will often be possible to find some subsets of parts in these tooling families which can be made at the same set-up. It is important in operation scheduling to make parts in the same tooling families on a machine, as far as possible, one after the other. It is unfortunate that the tooling families for different machines in a group do not always bring together the same sets of parts.

Some tooling families on a machine, may be different because they use different sets of tools, but they may be similar because they use many of the same tools.

Set-up time can also be saved by scheduling the work so that similar tooling families using many of the same tools, are loaded one after the other.

11.3.4 *Keeping bottleneck machines fully loaded*

In any processing department there will be one bottleneck machine, which is the most heavily loaded machine. In an expanding market, with standard process routes, this machine limits the maximum possible output of finished products from a factory. With GT there will still only be one bottleneck, and it will be found in one group only. Each group will, however, have its own most heavily loaded machine, which is its effective bottleneck.

It is essential in operation scheduling to keep these bottleneck machines fully occupied, and to ensure that none of their potential capacity is wasted. Components should be started on their first operations in a sequence which will keep the bottleneck machines fully loaded and ensure that there is a pool of work ready behind it, so that the risk of delay is minimized.

Sequencing for minimum set-up time, which loads parts in the same tooling families one after the other, is one way of maximizing the capacity of bottleneck machines.

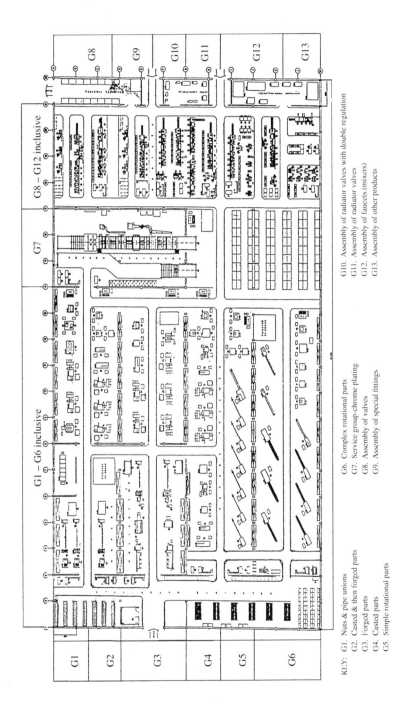

FIG. 11.3 A factory making bathroom and kitchen fittings.

KEY:
G1. Nuts & pipe unions
G2. Casted & then forged parts
G3. Forged parts
G4. Casted parts
G5. Simple rotational parts
G6. Complex rotational parts
G7. Service group-chrome plating
G8. Assembly of valves
G9. Assembly of special fittings
G10. Assembly of radiator valves with double regulation
G11. Assembly of radiator valves
G12. Assembly of faucets (mixers)
G13. Assembly of other products

11.3.5 *Scheduling for batch processes*

There are some intermediate operations for which several different parts must be accumulated into batches before processing. Examples can be found in heat treatment using batch-type furnaces. The early operations for parts requiring such processes must be started early in the week, to allow time to accumulate parts for the batch process, and to allow time for any following operations.

Figure 11.1 shows the case of a group which makes gears, twenty of which were hardened by the batch process of nitriding. The first machining operations on gears requiring nitriding are planned for completion in two lots, one by Tuesday evening and the second by Wednesday evening. The nitriding process is then done on Tuesday and Wednesday nights, leaving Thursday and Friday for the final operations.

11.3.6 *Scheduling to suit related processing stages*

It is sometimes possible to reduce throughout times and stocks by scheduling to eliminate a complete processing stage. Figure 11.3 shows a factory making bathroom and kitchen taps and related products. Raw-materials processing, component processing, electroplating, and assembly were originally treated as four separate manufacturing stages. As explained earlier in this book, die casting and hot pressing were later done in the same groups as machining, to eliminate one of the stages. Further studies showed that only 207 of the components which were produced required electroplating, but these came from nearly all the different component-processing groups.

It was found that if work started to flow into plating by the third hour on Monday morning, and the last parts for electroplating were finished and passed to the plating shop before the end of work on Wednesday, they could all be plated before the close of work on Friday. As already illustrated in Fig. 8.4, this eliminated another processing stage from the production-flow analysis (PFA) schedule.

11.4 Factors affecting the reliability of LSS

The starting sequence for first-operation machines, will depend on the nature of the groups (and, in particular, on the types of machine used), on the nature of the parts they make, and on the way in which manufacture has been divided into operations.

The reliability of the LSS method depends on a number of different factors, which are listed in Fig. 11.4. These factors can be ranked between 0, which indicates that LSS is very reliable, and 9, when LSS is more difficult to use.

11.4.1 *The number of parts per group and the number per period*

The first factor grades groups according to the number of different parts they make, and according to the quantity of each which are produced each period. If a company makes only a small number of products and most of the

	Factor	Reliable (0, 1, ...)	Difficult (... 8, 9)
1	Number of parts per group	Few. The same every period	Many. Different every period
2	Number of operations per part	Few. The same all parts	Many. Different for each part
3	Operation times	Short, all similar	Long or wide differences
4	Materials-flow system	Simple	Complex
5	Subcontracted operations	Nil, or first or last operation	Intermediate operations
6	Set-up tools per part	Simple, few tools per part	Complex, many tools per part
7	Set-up tool storage	By the machine, short SU	Central tool store, long SU
8	Bottleneck machines	First machine in route	Last machine or intermediate

FIG. 11.4 The reliability of LSS.

different parts in the group family are made every period (rank, 0 or 1), LSS will be more reliable than it will be if the company makes many different products, and the lists of parts made each period are widely different (ranked 8 or 9).

In practice, this is not a factor which limits the feasibility of using the LSS method. It only make it a little more complicated to control. If a company makes a range of similar products—say, all machine tools, or all pumps, or all valves, or all electric motors, or all internal-combustion engines—the proportion of different types of part in their parts lists will, as shown by research into component statistics, tend to be similar for all the products and variants they make. If the assembly groups are fully loaded every period, the load on the component-processing groups and their machines will also tend to be much the same every period.

11.4.2 *The number of operations per part*

The second factor in Fig. 11.4 grades groups according to the number of operations per part. If there are few operations per part, and the number is about the same for all parts (ranked 0 or 1), LSS will tend to be a very reliable method of scheduling. If there are many operations per part, or the number varies greatly (ranked 8 or 9), then LSS will be more difficult to use.

The number of operations per part largely determines the value of the throughput time. The most common case is one where a few *critical parts* have more than the average number of operations, and therefore they also have longer throughput times. One way of overcoming this problem is to give these parts special priority. Each container load of such parts carries a priority flag. When it is moved to the machine for the next operation, it goes to the head of the queue of unflagged containers.

11.4.3 *The operation time per operation*

The third factor grades groups according to the operation times for operations done in the group. Launch-sequence scheduling is more reliable if the operation times are short and if they are all approximately the same length (ranked 0, 1, etc.) than if the operation times are long or if they differ widely in length (ranked 8 or 9). Priority flags may again be useful in this latter case.

11.4.4 *The material-flow system*

Launch-sequence scheduling is most reliable when there is a simple MFS in the group (ranked 0 or 1), and it is more difficult to control when there is a complex MFS (ranked 8 or 9).

This is the key factor. If there is a simple MFS, this generally means that machines for different operations are normally used in the same sequence. Under these conditions, the sequence of loading the first-operation machines is more likely to be maintained on the following machines.

11.4.5 *Subcontracting*

It is always difficult to regulate the flow of parts to and from subcontractors for intermediate operations. Groups which require such operations should be ranked 9. Groups with no subcontracting operations can be ranked 0. Intermediate subcontracted operations should be eliminated by either bringing the work back into the group, or by buying the parts complete instead of making them.

Subcontracted operations which are the first operations on their routes can usually be accommodated. For example, if the first operation on a part is to flame-cut blanks from plate for machining, and if this is subcontracted, it is in effect only a specification of the form in which materials are to be purchased.

Subcontracted last operations can also be accommodated, but this generally requires an additional period and stage in the PBC ordering schedule. If, for example, electroplating is always a last operation, and it is subcontracted, an additional period for finishing may be needed in the ordering schedule to cover parts delivery to the subcontractor, electroplating, and the return of plated parts.

11.4.6 *Set-up—tools per part*

The time required for setting-up machine tools, is mainly a function of the number of tools per part, per operation, for those parts made in a group, and of the method used to store tools. Drilling machines, for example, are generally very simple to set-up, particularly if the number of drills per part is small. It should be possible to change the set-up on these machines in two or three minutes if the tools are stored in the group.

Turning centres on the other hand, with many tooling positions, each holding a different tool, have long set-up-time characteristics. To reduce set-up times and to increase the capacity in this case, use must be made of such as presetting and the sequencing of parts in the same tooling families for machining one after the other. Groups with mainly simple, easy to set-up, machines, will be ranked 0 or 1. Groups with more complex machines will be ranked 6, 7, 8, or 9.

11.4.7 *Set-up—tool storage*

Make-ready and presetting are major methods for reducing set-up times. The time required to make-ready the tooling for setting-up, or for presetting, depends mainly on the method used for tool storage. Storage in the group (ranked 1) gives the shortest set-up time. Storage in a central tool store, away from the group, tends, on the other hand, to increase set-up times.

11.4.8 *The position of the bottleneck in the route*

The position of the bottleneck machine in the process route also affects the reliability of LSS. If the bottleneck is the first machine in the route, the sequence of loading on this machine will be very easy to regulate. If, on the other hand, the bottleneck is used for operations which come late in the routeing sequence, it will be more difficult to ensure that parts in the same tooling family on the bottleneck are machined one after the other. In some cases it may be necessary to introduce supplementary controls to ensure that such parts are machined in sequence.

11.5 Planning the component launch sequence for a group

Figure 11.5 shows the *ranking*—defined in the previous section—for the seven GT groups in an engineering machine shop. This data shows that the problem of planning the launching sequence is different for each group, but that, at least in this case, and probably in the majority of GT groups, LSS is easy to plan and simple to control.

Consider groups 6 and 7 first. In these cases the ranking-index number (top-down) starts with a high number, (7) indicating many different parts, but after that it is small, indicating a simple case for LSS. This is because the number of operations per part is small; the MFS is simple (indicating a common sequence of machine usage). The set-up times on all machines are short (due to the few tool types and the tool storage by the machines); the bottleneck is fictitious (because the most heavily loaded machine in both cases is a machine for which there are three of the type in the group: two are fully loaded, and the third is underloaded); and, finally, because there are no intermediate subcontracted operations. In these circumstances almost any sequence of loading can be used without affecting the group capacity, or the completion of the period list order by the due date.

Groups 1, 2, and 4 are similar, but they contain machines with longer natural set-up times due to the large number of different tools per part. Because, however, there are still few operations per part, because there is a simple MFS with

Factor		Group number and name						
		1 Cases	2 Shafts	3 Gears	4 Bar	5 Pyramidal	6 Turning, milling, drilling cast iron	7 Turning, milling, drilling steel
1	Parts per group	3	2	3	8	2	7	7
2	Operations per part	4	4	6	3	8	3	3
3	Operation times per operation	6	7	7	2	7	4	4
4	Materials-flow system	2	3	3	2	5	3	3
5	Subcontracted	0	0	0	0	1	0	0
6	SU tools per part	6	6	4	7	6	4	4
7	SU tool store	2	1	1	2	2	2	2
8	Bottleneck in route	2	3	7	1	8	1	1
	Total	25	26	31	25	39	24	24

FIG. 11.5 The ranking of machine groups for LSS.

a common machine-usage sequence, because the bottleneck machine is used early in the routes, and because there are no intermediate subcontracted operations, it is not difficult to plan a launching sequence which will give the optimum loading sequence on those machines with long set-up-time characteristics.

Groups 3 and 5 are the most difficult for LSS. They have many operations per part and long operation times; their bottleneck machines are used for intermediate operations, and they are special machines (there is only one of that type). It is probable that some of the parts in these groups will have much longer throughput times than the others, and the groups will have to work shifts and the critical parts be given scheduling priority. On the other hand, it will be important to machine parts in the same tooling families one after the other, to conserve capacity and to reduce throughput times, particularly on the bottleneck machines.

11.6 Flexibility with LSS

One of the main problems with the traditional methods of operation scheduling used with process organization and multicycle ordering is that they are very inflexible. Any change that is necessary in the planned schedule tends to require a complete replanning of the total operation schedule. In practice, when this becomes necessary, the machine shop cannot stop work and the manager of the workshop is told to do the best he can.

With GT and PBC, the schedules cover single groups and they are therefore much shorter and simpler. In addition, GT tends to bring together larger tooling families. Changes in the loading sequence inside the same tooling families can be accommodated with little increase in the set-up time.

The foreman in a group who is intimately acquainted with his own small group of machines and small family of parts is in a much better position to cope with scheduling problems than is the manager of a traditional department with process organization and multicycle ordering.

11.7 Summary

Operation scheduling with GT and PBC is much simpler and more reliable than is the case with process organization and multicycle ordering systems such as stock control and MRP.

With GT and PBC, LSS can be used, in which the sequence of loading parts each period, on the machines for first operations, is planned first. In most cases, if there is sufficient capacity, this plan coupled with strict queuing discipline on the machines for following operations will make it possible to complete all parts by the end of the period.

PART 3

PBC IN PRACTICE

12

PBC FOR IMPLOSIVE INDUSTRIES

12.1 Introduction

Implosive industries are those which convert a small number of material varieties into a large number of different products. Typical examples include cast-iron foundries, which start with pig iron and scrap iron and produce large numbers of different castings; potteries, which make teapots, plates, cups and saucers, and other products from clay; glass works, which make many different types and sizes of bottles, vases, jugs and drinking glasses, mainly from sand; and factories making many varieties of decorative laminates, starting with paper, resin, coloured or patterned sheets, and sheets of transparent melamine plastic.

These implosive industries tend to pose similar production control (PC) problems in all cases. This chapter examines, first, the design of a Period Batch Control (PBC) system for the control of production in a factory making decorative laminates. It then briefly examines, as a second example, a factory making decorated glassware.

12.2 The product

Decorative laminates are sheets of a hard, wear-resistant, composite material, with an attractive coloured or patterned surface, which are used mainly in the building and furniture industries, and by do-it-yourself enthusiasts.

They consist of several layers of paper, impregnated with a resinous compound, topped by a coloured or illustrated sheet, which is protected by a hard transparent sheet of melamine plastic. Several of these piles are stacked on top of each other. The stack is consolidated in a press, under a controlled pressure and temperature, to form hard, thin, decorative sheets which are normally glued onto materials such as wood or chipboard, to provide a hard, wear-resisting, decorative surface.

The decorative sheets are in plain colours—white, black, cream, red, blue, green, and yellow etc.—or in wood-grain patterns—oak, ash, teak, etc.—or they may take the form of paintings, or drawings specially commissioned by a particular customer. In the company considered here, there is a special art department which designs the decorative sheets, and there is a printing department which prints them ready for production. Other companies may subcontract this work.

Sales to builders, furniture manufacturers, and the general public are made through a distribution chain of wholesalers and retailers. The wholesaler's orders to the factory normally include different quantities of a range of different colours, patterns, and sheet sizes, which have to be *picked* in sets from the stores racks to make up the orders.

12.3 The method of manufacture

The method of manufacture is illustrated in Fig. 12.1. The resin impregnation is performed by drawing the paper through baths of a heated liquid resin. The sheets are cut to length by a flying cutter, stacked to the required thickness, and then capped with a decorative sheet and a melamine cover.

The stacks are piled, consolidated in a hydraulic press under controlled pressure and temperature, and then cooled. Next, they are trimmed by a guillotine to remove edge imperfections and they are cut to suit any special size requirements given in the orders.

In the factory under consideration, there are four of these impregnation, stacking, pressing, and guillotining lines, that is, there are four Group Technology (GT) groups. Each group is independent and is held responsible for its own quality and for completion of orders by due date.

12.4 The annual programme

Annual sales, production, and stock programmes are required, and they are made in the way described in Chapter 6. They are needed as the basis for planning financial controls and cash flow, as a guide for the purchasing of paper and resin, and for the long-term planning of changes in capacity.

The products considered in the programmes are all the individual different standard colours and patterns, plus all special designs, which are treated for control purposes as a single category. The unit of measurement is the consolidated *sheet* as delivered from the press, and the unit of time is the period, or the working week.

Records are maintained of actual and moving annual total, (MAT) sales, and of seasonal variations (see Figs 6.5 and 6.6). The cumulative output figures each period are calculated for the annual programme and also for the series of short-term programmes to be described later. Variances are calculated, and they are considered at the weekly programme meetings. If the difference between them becomes unreasonably large, the annual programme is corrected and the financial budget, cash-flow plans, and purchase forecasts are corrected accordingly.

12.5 The short-term programmes

The short-term programmes which form the foundation for the PBC system are based, in the case of standard laminates, on the accumulation of the sales

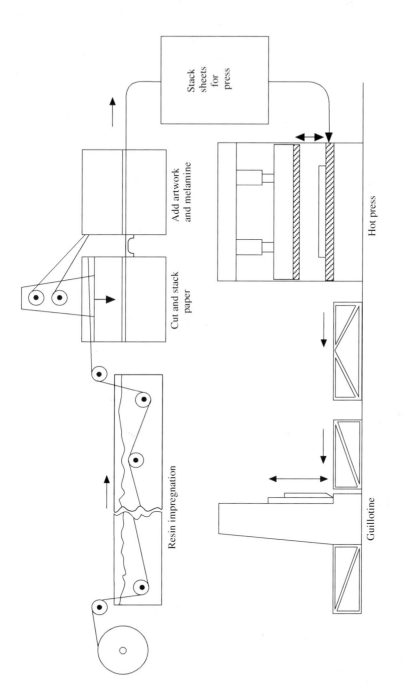

FIG. 12.1 The manufacture of decorative laminates.

orders received each week. In the case of jobbing orders (special designs), capacity is reserved for the work. As soon as the special decorative sheets are ready, the orders are filed in a *ready file* which is studied weekly at the programme meeting. These orders are added to the short-term programmes, which consist mainly of standard sheets, at the discretion of the meeting (see Fig. 12.2).

As is the case with most implosive systems, the products produced are components, which are sheets in this case. They are, in a sense, also assemblies, but since they all contain the same parts—other than the decorative sheets—there is no need for a special ordering system in addition to the programme, except for the programme needed to order paper, resins, melamine, and standard decorative sheets. These needs will be examined later.

Based on the statistical trend of orders received, a mean *demand/capacity level* per week is fixed (C, giving the number of sheets to be produced per week).

(a) Programming schedule

Week number	1	2	3	4	5	6
Cycle 1	Ac ● M	S				
Cycle 2		Ac ● M	S			
Cycle 3			Ac ● M	S		

Ac = Accumulate orders
● = Programme meeting
M = Make
S = Distribution

(b) Short-term programme

Short-term programme			Group number: 3				Period number: 26
Item number	Decorative laminate		Sheets				Programme
	Number	Description	Ac	Stock	MM	+ smooth	
1		White	40	14	26	10	36
2		Yellow	15	2	13	10	23
3		Green	50	10	40	10	50
4		Blue	40	9	31	10	41
5		Orange	15	6	9	10	19
6		Red	20	1	19	10	29
7		Black	20	4	16	10	26
8		Oak	10	3	7	5	12
9		Teak	12	1	11	5	16
10		Walnut	10	–	10	5	15
11		Beech	15	1	14	5	19
12		Special artwork	4	–	4	–	4
Totals			251	51	200	90	290

Key: Ac = accumulate; M/M = must make, + smooth = add smoothing stock.
Note, the capacity, C, is 290 sheets per week.

FIG. 12.2 A short-term programme for decorative laminates.

This capacity level is checked against the load each week at the programme meeting, and it is changed when it differs so much from the load that it cannot be covered by changes in overtime or by smoothing.

The accumulation of sales orders, Q, each week, ceases on Friday afternoon and it then starts again for the next week. Receipts and issues of laminates to and from the stores stop on Friday afternoon at five p.m. and they start again on Saturday morning. The stock of all standard items in the store, S, is physically checked on Friday afternoon.

The quantity to be ordered for manufacture of each standard design of sheet, is calculated on Friday afternoon each week, just before the programme meeting, as follows:

1. Find the *must-make quantity* (MM) for each product—or the orders accumulated for each product in the week, Q, less the stock for each product, S, that is, MM = $Q - S$ sheets.
2. Find the total number of standard sheets required for all products = Σ MM sheets.
3. Add any *special design orders*, SD, which are ready to run, and for which there is capacity, to find the total T.
4. Compare the total load T, with the capacity, C,
   (a) If $T > C$; make C sheets and draw on stocks, S, to supply the additional $T - C$ sheets needed.
   (b) If $T < C$; make in addition $C - T$ extra, selected, high-usage, standard sheets for stock to use all the capacity.

The period chosen for short-term programming was one week, the same as that used for the annual programme. It is noted that the only time constraint involved in this case is the capacity constraint (see Section 9.2). The process is continuous, so there is no problem with the throughput time, and the set-up times to change from one decorative sheet to another are negligible.

12.6 Ordering materials

The small number of different materials which have to be ordered to support manufacture are:

(1) base paper in rolls;
(2) resins;
(3) melamine sheets;
(4) paper for printing;
(5) printing inks.

All of these materials are ordered from outside suppliers. The requirements of the first three are directly proportional to the chosen demand/capacity level, C.

In addition, special designs have to be ordered from the artwork department and standard designs and coloured sheets have to be ordered from the printing department, which in turn requisition their own materials (see (4) and (5) above).

12.6.1 *Ordering materials from suppliers*

The call-off method of purchasing is used for ordering materials, with an insurance buffer stock, held in the factory, of two-week supply. Materials are issued daily from stock to manufacturing as needed. The quantities issued each week are called-off for replacement from suppliers on Friday each week for delivery by Friday in the following week (see Fig. 12.3).

The actual stock in the material stores varies between two and seven working days supply. Most material issues are made on Monday morning each week. The stock of each item in the stores is checked immediately after every issue.

12.6.2 *Ordering special designs*

The ordering of special designs from the artwork department is less systematic than the ordering of materials. The artwork department is often involved in consultation with the customer before the contract is placed, and a lot of the design work may have already been done before that point is reached.

Not all design work leads to sales contracts. The time spent by artists on different enquiries and orders is carefully recorded, and care is taken to see that

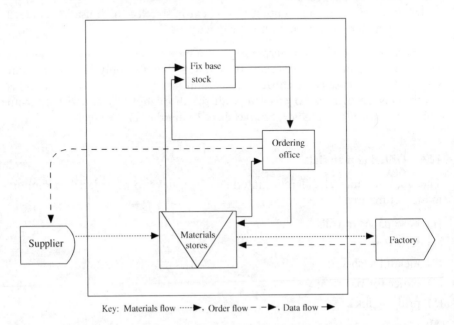

Key: Materials flow ·····▶, Order flow − −▶, Data flow ─▶

FIG. 12.3 Ordering basic materials.

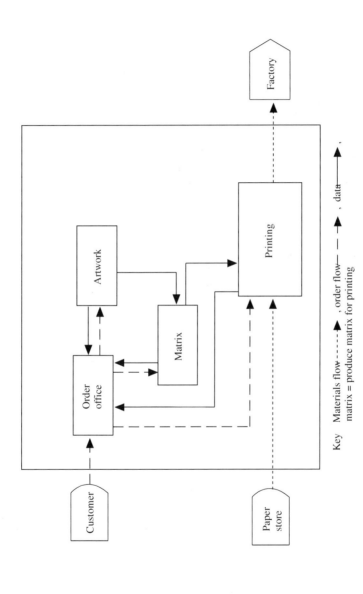

FIG. 12.4 The ordering or the printing group, in a decorative laminates factory.

all expenditure in the artwork department is recovered by the prices charged for actual sales.

12.6.3 *Ordering the printing of decorative sheets*

Orders on the printing department for decorative sheets are issued by the PC department. For standard colours and patterns, the PBC method of ordering is used, supported by delivery-buffer stocks (see Fig. 12.4).

There is a problem with *colour matching* in the case of single-colour sheets. It can be difficult to achieve a match between the colours of different batches which is true in both natural and artificial light. For this reason, in the case of large orders where colour matching is important, it may be desirable to order decorative sheets all of the same colour, to be run from the same roll of paper, with the same ink mix, on the same continuous printing run.

The need to differentiate between different batches of the same product greatly complicates production. A continuous effort is made, therefore, to improve colour matching.

12.7 Ordering in other implosive industries

Similar methods can be used in other implosive industries, such as foundries, potteries, glass works, and spinning mills. The key step is planning the short-term programming system, which regulates the manufacture of all the different products produced.

Since, by definition, the implosive industries use only a small number of different material items, their ordering is generally simple to plan and control. In the case of decorative laminates, few basic materials are needed—paper, resin, and melamine—but the problem is complicated by the need to order the colour and pattern sheets. The introduction of PBC in other implosive industries is generally simpler.

Because there is normally an element of line flow of materials in most implosive industries, the throughput time is seldom a serious constraint. Where the set-up time is a constraint, it is generally simple to overcome.

12.7.1 *Ordering in a decorated-glass factory*

As a second example of PBC in an implosive industry, a factory making glass objects, with engraved or etched designs on them, will be considered. The processes of manufacture are illustrated in Fig. 12.5.

The factory is divided into four GT groups, making:

group 1, flower vases;

group 2, bowls;

group 3, drinking glasses;

group 4, jugs.

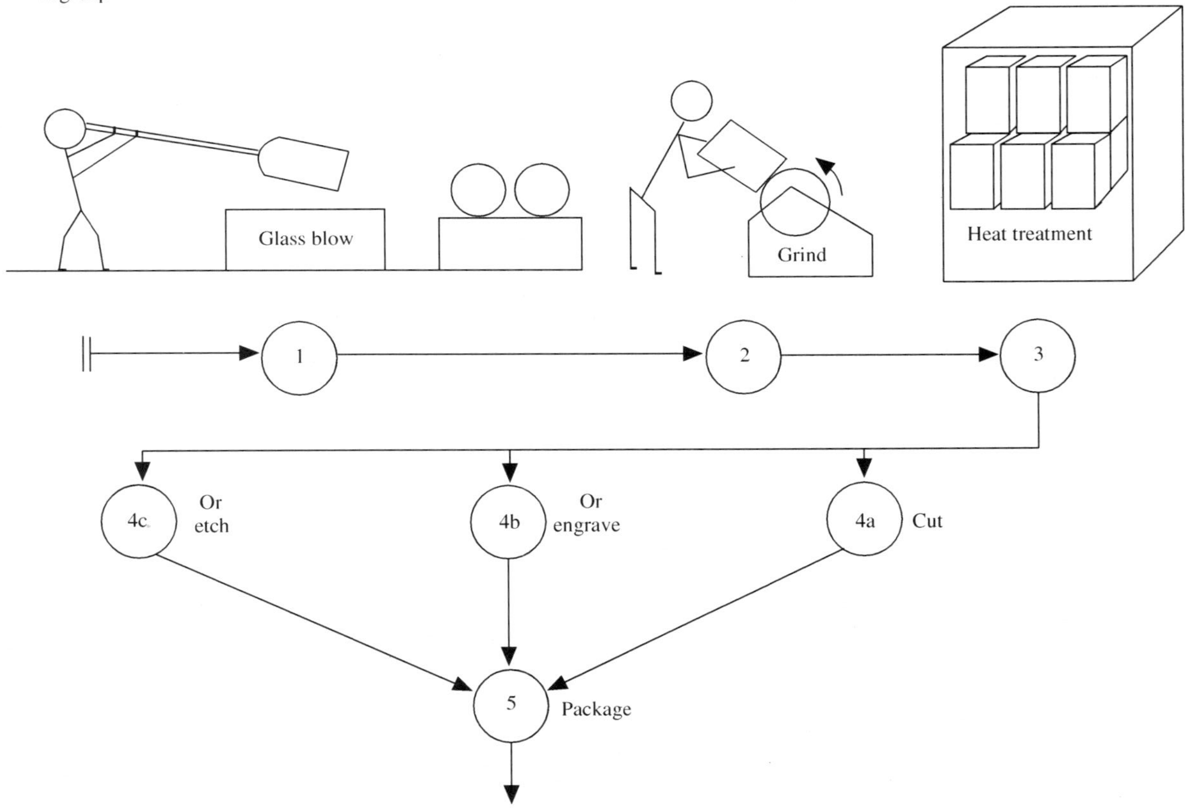

FIG. 12.5 The manufacture of decorated glassware.

The prime process in each group is glass blowing. The glass blowers blow and shape the various products to established designs, and then cut them from the end of the blow pipe. They are then ground to remove sharp edges and heat treated to toughen the glass.

They are decorated with a grinding wheel (cut glass); or an engraving tool; or by coating the object with wax, drawing the design with a needle, and then etching in acid.

Some products are made to special designs, but the majority of products are made for stock, to a catalogue of designs. A standard stock level is fixed for each product. The standard period is one week and the standard programming schedule covers four weeks, allowing one week for the accumulation of orders; one week for glass blowing, grinding, and heat treatment; one week for decoration; and one week for packing and delivery to stores.

Orders are delivered to customers from stock on the day following receipt of their orders. The orders received each week are accumulated for each group. Subject to smoothing to save capacity, these quantities are ordered on the groups, to replace the products despatched to customers from stock.

12.8 Summary

Period Batch Control is particularly well-suited for PC in the implosive industries because materials flow, in this case, is easily regulated by the series of short-term programmes. Because the implosive industries make many varieties of product from a few varieties of material, the ordering of materials is very simple.

In practice, the time constraints which must be eliminated to use PBC are easier to eliminate in the implosive industries than is the case in other types of industry.

The capacity and the load tend to be a function of the total number of items produced, or of their weight, and they are little affected by the design mix. A measure of line flow is generally an important characteristic of the materials-flow systems in the implosive industries. The simplicity of their materials flow systems means that the throughput time is seldom a problem. Finally, the set-up time is, again, seldom a major problem in this type of industry.

13

PBC IN EXPLOSIVE INDUSTRIES

13.1 Introduction

Explosive industries are those industries which convert a large number of material items into a much smaller number of finished products. The complexity of the manufacturing systems in the explosive industries depends on:

(1) the design of the product(s);
(2) the market for the product(s);
(3) the method of manufacture;
(4) the type of organization—process or product organisation.

This chapter studies factories which make assembled products, comprising a number of different parts, either bought or made in house, which are assembled together to make products. It is assumed that these companies use product organization, (Group Technology GT) rather than process organization. Because GT is generally possible and always more efficient than process organisation.

Examples of industries which are in this category include, at the simpler end of the range, factories making centrifugal pumps, hydraulic valves and taps, some types of agricultural machinery, and some types of electrical-switch gear. The more complex end of the range of assembled products, includes aircraft and machine tools. This chapter first uses, as an example, the case of a factory making a range of agricultural machines. Later, the more complex case of Period Batch Control (PBC) in a machine-tool factory is described.

13.2 The range of products produced

The range of products produced in the agricultural machinery factory consists of tractor-mounted tilling, seeding, planting, hoeing, and harvesting machines designed for small farms. The products in the range are changed from time to time to suit market needs, but typical examples include ploughs, disc harrows, rollers, seed drills, mowers, transplanters, steerable hoes, swath turners, hay rakes, and potato diggers.

Product sales are highly seasonal, and the range has been designed to provide a seasonally balanced set of products each of which can be produced intermittently, as required throughout the year, with a minimum stock of materials, parts, and finished products. As there is a short sales period for each product, however, and as the market requires sales delivery from stock, some products have to be made for stock in anticipation of future sales at the beginning of each product season.

13.3 The materials-flow system

The materials flow system (MFS) of the enterprise is illustrated in Fig. 13.1. Seven GT component-processing groups (C1–C7) make components for two assembly groups and a spare-parts store. A centralized painting group (P) paints the finished products, and it also paints some parts which are sold as

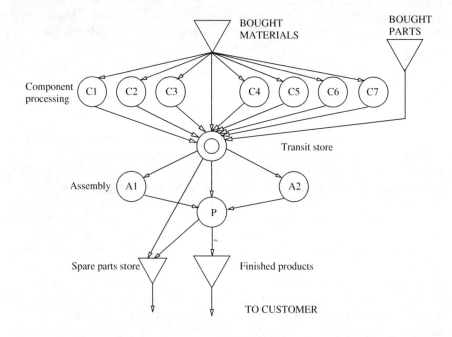

Group		Group	
C1	Cast-iron parts	C6	Cold forge and weld (2)
C2	Steel-turned parts	C7	Sheet-metal work
C3	Small rotational parts	A1	Assembly (1)
C4	Pyramidical parts	A2	Assembly (2)
C5	Cold-forge and weld (1)	P	Painting

FIG. 13.1 An MFS for agricultural machines.

spare parts in the painted condition. The distribution of labour, machine tools, parts, and products between the manufacturing groups is given in Table 13.1.

All the groups complete all the parts or products they make, at the stage where they are installed and they have all the facilities inside the groups that they need to do so. There is one *service group*—painting. There are no subcontracted intermediate operations.

Table 13.1 Group size

Item	Group number										Totals
	C1	C2	C3	C4	C5	C6	C7	A1	A2	P	
People	10	12	15	14	7	11	18	14	13	8	122
Machines	15	19	22	25	13	18	32	–	–	–	144
Parts	120	150	316	83	96	117	183	–	–	–	1065
Products	–	–	–	–	–	–	–	11	16	(27)	27

In addition to the 1065 made parts, there are 796 different bought parts. This figure includes 134 parts which could be made in the factory, but have been subcontracted complete, to reduce the load on heavily loaded machine tools.

The two cold-forging and welding groups (C5 and C6) make parts—using flame cutting from steel plate and simple machining—weld some of them together to form subassemblies, and shot-blast and prime-paint the assemblies and other components they produce. The sheet metal group (C7) also makes some welded assemblies and prime-paints some of its output. Two of the machining groups (C2 and C4) contain heat-treatment equipment for hardening.

The capacity level is set to meet the present market-demand level. At this level, it is always possible to complete all the products or parts programmed for manufacture each period in a group, in a throughput time of less than one week. Tooling analysis (TA)—part of production-flow analysis (PFA)—has been used to find *tooling families* in the groups, of parts which can be machined with the same, or very similar, set-ups on the more heavily loaded machines. Provided that parts in these tooling families are always machined one after the other (sequencing), and that tools are ready for the next job before the previous job is completed, the change to weekly ordering can be accommodated, without a loss in the total group capacity due to the increased number of set-ups.

All output from groups, and the input of bought parts on call-off from suppliers, is routed to a *transit store*, where it is counted and is assigned to either an assembly group or to the spare-parts store. The assembly parts are accumulated in the transit store until the end of each week, when they are moved together to the appropriate assembly group.

13.4 Planning PBC for the factory

The PBC system for this factory was planned in nine main stages:

stage 1, select the standard period;

stage 2, select the ordering schedule;

stage 3, fix the capacity level;

stage 4, plan the ordering standards;

stage 5, plan the period sales programme;

stage 6, plan the period production programme;

stage 7, plan the period orders and the load summary;

stage 8, plan the call-off of purchases;

stage 9, plan the feedback controls.

13.4.1 *Selecting the period*

Because all parts could be completed in the groups within a throughput time of one week, the week was selected as the chosen period for PBC. As far as manufacturing was concerned, there were very few critical parts, and a shorter period might have been possible. The limiting factor was finding a period, giving a lead time for call-off which was acceptable to the suppliers.

13.4.2 *Selecting the ordering schedule*

The four-period ordering schedule chosen for the factory is illustrated in Fig. 13.2. A programme meeting is held on Friday afternoon each week to approve the sales and production programmes for the coming cycle.

Call-off instructions are despatched to suppliers on Friday evening mainly through direct computer links, giving a one-week lead time for materials, and a two-week lead time for the call-off of bought parts. Week 1 is reserved for obtaining material deliveries from suppliers. Week 2 is used for the manufacture of components in groups C1–C7 (and P) and for the receipt of bought

Week number	1	2	3	4	5	6
Cycle 1	●	O	M	A	S / P	
Cycle 2		●	O	M	A	S / P
Cycle 3			●	O	M	A	S / P

Key:

● = Programme meeting

O = Obtain materials

M = Make parts

A = Assemble/paint

S = Despatch

P = Painting

Assembly is a continuous process. Painting starts as soon as the first product is assembled in week 3, and it is completed for the last product early in week 4. When dry, the finished products are moved to finished stores ready for delivery to customers.

FIG. 13.2 An ordering schedule for agricultural machines.

parts on call-off from suppliers, allowing 2 weeks lead time. Week 3 is used for assembly, and Week 4 is reserved for the delivery of finished products to customers (usually wholesalers).

13.4.3 Fixing the capacity level

The annual production programme shows the start assembly date and the forecast smoothed capacity requirement each period for each product.

A statistical control of the Sales orders received is used to correct the assembly capacity level when necessary. Assembly capacity is reviewed at the programme meeting each week.

Component processing load is measured each week for all the machines in each group, based on the ordering standards and on the period production programme. Component-processing capacity is fixed for all machines, to meet the required load. The capacity on these machines is adjusted when necessary, by adding overtime, by taking on new employees by shift working, or by sub-contracting parts manufacture, to match changes in demand.

13.4.4 Planning the ordering standards

With GT and PBC, one of each product type in the period production programme imposes a standard:

(1) requirement for parts from each group;
(2) requirement for materials for each group;
(3) load in machine hours on each machine in each group;
(4) requirement for bought parts to be called off for each assembly group;
(5) load in man-hours for each assembly group.

These ordering standards are illustrated in Fig. 13.3. Once the production programme is known, simple multiplication of these standards by the weekly production programme quantities, provides the group-list orders, the purchasing call-off notes and the load summaries needed for production control (PC).

13.4.5 Planning the period sales programme

The sales cycle for each type of product is subject to seasonal demand characteristics. It may only be possible to sell some products in a limited season of two or three months.

The sales of each product depends partly on having the product on display in wholesalers' and retailers' show rooms before the season starts, partly on exhibiting them at agricultural shows, and partly on advertising,

The annual sales programme is based partly on an <u>estimate of these special requirements,</u> partly on statistical forecasting, and <u>partly on market research.</u>

The period sales programme for each product starts at the beginning of the sales season for a product. This is shown in the annual programme with the

Made parts		Product	Group number
Part number	Description	Quantity of each part per product	

Materials		Product	Group number
Code	Description	Quantity of each part per product	

Load in machine-hours		Product	Group number
Machine code	Description	Machine-hours per product	

Bought parts call-off		Product	Group number
Part number	Description	Quantity of each part per product	Supplier

Load in man-hours for each assembly		Product	Group number
Code number	Assembly	Man-hours per product	

FIG. 13.3 Ordering standards.

issue of a small smoothing stock and the creation of priming orders, to cover the needs of show rooms and agricultural shows. As the season progresses, the short-term (period) sales programmes include all new sales orders received, each period. The short-term production programme shows these figures after smoothing. The capacity level used for smoothing starts at the annual programme level, but is amended at the period programme meetings, as necessary to end the product season with minimum stocks.

13.4.6 Planning the period production programme

The annual production programme attempts to supply the requirements of the annual sales programme and, at the same time, to smooth any residual seasonal variations in the load which have not been eliminated by the seasonally balanced product range. This is done by planning manufacture of some products for stock when the forecast demand is less than the capacity level, and using these stocks to supplement the programmed output later, when it is above the capacity level.

The short-term, or period production, programme attempts to supply the requirements of the short-term or period sales programme, but at the same time it attempts to smooth production, in this case to compensate for random variations in sales demand. Smoothing is based on differences between the period sales programme and the capacity. At the end of each product's sales season, it may be necessary to modify the smoothing rules to provide for additional sales, and to ensure that one ends each product sales season with zero stock. When the short-term programmes are issued each week, the forecast of the cumulative total output, to that date, in the annual programme is compared with the cumulative total output in the series of short-term programmes for the same period. If the variance between these values becomes excessive, the annual programme is revised. In this case, any budgets, cash-flow forecasts, or purchasing-requirement forecasts which were based on the old annual programme must also be revised.

13.4.7 Planning the period workshop orders and load summaries

The short-term production programme provides the authority for assembly in the two assembly groups. The list orders for each component-processing group, the materials supply lists and load summaries for the same groups, the load summaries for the assembly groups, and the call-off notes for suppliers are all found by multiplying the product output shown in the short-term production programme by the ordering standards illustrated in Fig. 13.3.

In addition, the above orders will have to be adjusted to cover any additional parts needed for sale as spare parts. To make this adjustment, ordering standards are also prepared for each part, showing the group source, the materials requirement, and the load on machines.

13.4.8 Planning the call-off of purchases

Call-off instructions for suppliers are again generated by multiplying the numbers of products shown in the short-term production programme by the ordering standards for bought parts. Call-off from most suppliers is by a direct computer link.

13.4.9 *Planning the feed back control*

In the present case, the following indicators were monitored:

(1) products overdue for completion at the period end,
(2) parts overdue for assembly, at the period end,

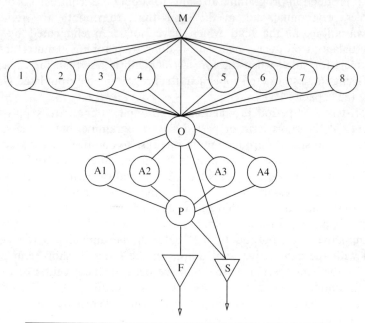

Group		Group	
Number	Name	Number	Name
M	Materials store	8	Gears
1	Heavy beds	O	Transit store
2	Large C1 parts	A1	Assembly 1
3	C1 turned	A2	Assembly 2
4	Pyramidic parts	A3	Assembly 3
5	Small rotational parts	A4	Assembly 4
6	Steel turned	P	Paint
7	Non-ferrous	F	Finished product
		S	Spare parts

FIG. 13.4 A flow system for a machine-tool factory.

(3) parts produced in excess of orders, at the end of each period,
(4) quality rejects per million parts produced at the period end,
(5) absentees man-hours per week,
(6) machine down time, in machine-hours per week.

The assembly groups report on the first indicator. The second and third indicators are monitored by the transit store. The fourth indicator is reported by the component groups, and the fifth and sixth indicators are again reported by the groups.

13.5 PBC in a machine-tool factory

A group layout of a machine-tool factory, making a range of grinding machines, has already been illustrated in Fig. 2.12. The MFS and standard ordering schedule for PBC for this factory are illustrated in Figs 13.4 and 13.5.

The MFS is similar in complexity to that for the agricultural-machinery factory. A standard period of one week was again selected. Although the machine-tool parts had more operations and longer operation times than those for the agricultural machinery, there was a lot of spare capacity on most machines, and the period requirement for most parts could be completed in one period. A limited second shift found sufficient extra capacity to eliminate the throughput-time problem for the remaining parts.

Some reduction in the set-up times was needed to save capacity and to make it possible to work with the higher set-up frequencies imposed by the use of

Week No	1	2	3	4	5	6	7	8
Cycle 1	Ac	O	M	A	S			
Cycle 2		Ac	O	M	A	S		
Cycle 3			Ac	O	M	A	S	
Cycle 4				Ac	O	M	A	S

Key:

Ac = Accumulate orders
O = Obtain materials
M = Make parts
A = Assemble
S = Deliver to customer

FIG. 13.5 An ordering schedule for PBC in a machine-tool factory.

one-week periods. Moving the tool storage into the groups, coupled with efficient make-ready, pre-setting and sequencing procedures, gained most of the savings required.

It can be seen that the sales programme for week 5 in each cycle was found by accumulating the orders received in week 1. The assembly programme for week 4 covered the needs of the sales programme for the cycle, but it was smoothed to save capacity and to give an even rate of production.

Explosion from the assembly programme, based on ordering standards like those shown in Fig. 13.3, produced the list orders for each group and call-off instructions for suppliers.

13.6 The advantages of PBC

The advantages of GT have already been listed in Fig. 2.5. Most of these advantages are reinforced by the introduction of PBC. In addition, PBC gives a number of extra advantages in its own right, of which the following are important in the present case:

1. It eliminates the costly and time-consuming process of kitting-out for assembly. Parts are made each period in the exact quantities needed for assembly in the next period.

2. It eliminates the need for a finished-parts store—other than that needed for spare parts—for the same reason.

3. It eliminates the surge effect, because it is a single-cycle system (see Fig. 1.7).

4. Loading, or comparing the load and the capacity, is much simpler with GT and PBC than it is with traditional process organization and multicycle ordering, because the load checked is always limited to a single period.

13.7 The problems with PBC

The problems which are likely to be experienced when introducing PBC include:

1. The programming of a seasonally balanced range of products is very difficult. There is an attempt to achieve both no lost sales and zero stock at the end of the sales season for each product. The demand for the products can varies not only with the weather, but also with the seasons, with the economic climate, and with the competition. Success depends partly on *flair*.

2. The load in machine-hours and man-hours imposed by the short-term programme on the groups must not be greater than the capacity available. It is not difficult to check the load against the capacity with GT and PBC, and these checks should be made every week.

3. Even with complex products, with PBC, GT, and a good plant layout, there should be no difficulty in completing all the parts inside the allowed throughput time of one week—168 hours.

4. The need to reduce setting-up times in order to conserve capacity will probably be the main problem. First, it is necessary to discover which machines have long set-up times, where a reduction in the set-up time is needed. Tooling analysis (TA), (part of PFA)—should be used to find tooling families of parts which can be made at the same set-up, and to find the sequence for loading parts on each machine to minimize the set-up time. The storage of tools in groups to reduce the *make-ready time* for set-up, the provision of tooling trollies (used to hold sets of tools ready for the next set-up) and the use of presetting are other relatively inexpensive methods for reducing set-up times.

5. Finally, the introduction of the call-off method for purchasing will involve visits and negotiations with selected suppliers. This is seldom as difficult to arrange as expected but it takes a long time to introduce for all purchased items.

13.8 Summary

Explosive industries are those which convert a large number of material items into a much smaller number of finished products. This chapter took as examples an engineering company which makes a range of agricultural machinery and a machine-tool factory. Both companies had already been converted to Group Technology. This chapter studies the design of Period Batch Control systems to regulate the flow of materials through the factories.

The agricultural industry is subject to major seasonal variations, so some degree of manufacture for stock, in anticipation of future orders, is unavoidable. A smoothing system based on differences between the period sales programme and the planned capacity, was used to generate these stocks. In the present case, the investment in stocks has been reduced by the introduction of a seasonally balanced range of products, and by the introduction of PBC with one-week periods. The four-week ordering schedule allows a two-week lead time for the call-off of bought parts, and one week for bought materials.

Contrary to normal expectations, it was simpler to introduce PBC in the machine-tool factory than in the agricultural-machinery factory, because in this case there was much less seasonal variation in demand. Machine-tool manufacture is, however, more technologically demanding, and the main problems in this case were to reduce the throughput times and the set-up times for a small number of critical parts.

14

PBC IN PROCESS AND SQUARE INDUSTRIES

14.1 Introduction

Process industries are those industries in which a small number of material varieties are converted into an equally small number of saleable products. Typical examples can be found in cement factories, in the treatment of mineral ores at a mine head, with many simple chemical products, and with many food and drink products (see Fig. 1.3).

Square industries are those industries in which a large number of material varieties are converted into an equal number of product varieties. The materials are generally supplied by the customer and they are returned after processing. Examples can be found in yarn and/cloth dyeing; in cloth-finishing mills; in laundries and dry-cleaning works; and in factories doing heat treatment, painting and electroplating, for example, as a service to other factories.

This chapter looks first at the process industries. These can vary in the seasonality of material input and of product output, in the seasonality of product demand from customers, in the delivery lead time acceptable to the customer, and in the shelf-life of the products produced. The first example considered is farm products, such as wool and potatoes, with highly seasonal production characteristics, secondly an examination is made of products with a much more even rate of output, such as cement, ores from a mine, and many chemicals. Finally a study is made of a food industry—pork-pies manufacture—in which the product has a short shelf-life.

This chapter ends with a brief study of square industries, such as the dyeing of yarns, the finishing of cloth, and services such as painting and electroplating for manufacturing factories.

14.2 Fixing the capacity and Output levels

Because, by definition, factories in process industries produce a very small number of different products, their output rate can be based on statistical forecasting with more accuracy than in other industries. If the market for the product is highly seasonal, *moving annual totals* (MATs) can be used to find the trend, coupled with period sales analysis to find the percentage of annual sales normally sold each period. If the market for the product is regular (that

(a) Moving annual totals (MAT)

MAT								Product A					1992
1991 Week No	1	2	3	4	5	6	7	8	9	10	11	12	13	Total
Orders	10	9	12	14	17	15	19	16	16	20	21	23	20	203 –
1991 Week number	14	15	16	17	18	19	20	21	22	23	24	25	26	
Orders	21	24	23	19	21	23	20	22	23	22	23	20	24	285 488
1991 Week number	27	28	29	30	31	32	33	34	35	36	37	38	39	
Orders	25	27	23	25	26	22	22	22	19	20	20	21	20	292 780
1991 Week number	40	41	42	43	44	45	46	47	48	49	50	51	52	
Orders	19	20	17	19	18	16	16	16	18	15	10	9	10	203 983
1992 Week number	1	2	3	4	5	6	7	8	9	10	11	12	13	
Orders	10	10	11	14	15	14	16	15	17	19	22	22	21	206
MAT	983	984	983	983	981	980	977	976	977	976	977	976	977	

(b) Four-week moving totals (MT)

Four-week moving average								Product A						Year: 1992	
Week Number	1	2	3	4	5	6	7	8	9	10	11	12	13	14	15	16
Orders	10	9	12	14	17	15	19	16	16	20	21	23	20	21	24	23
Four-week MT				45	52	58	65	67	66	71	73	80	84	85	88	88
Week number	17	18	19	20	21	22	23	24	25	26	27	28	29	30	31	32
Orders	19	21	23	20	22	23	22	23	20	24	25	27	23	25	26	22
Four-week MT	87	87	86	83	86	88	87	90	88	89	92	96	99	100	101	96

FIG. 14.1 The trend of orders received.

is, without strong seasonal variations), shorter period totals, or averages, will be more accurate in following the trends (see Fig. 14.1).

Knowing the planned output rate, and also knowing the production throughput time, it is simple to calculate the materials input schedule and to arrange for material deliveries to this schedule which will be sufficient to support the planned output. In the past, many factories in the process industries operated in this simple way. The only problems are that the total level of demand tends to vary with time, and that there will normally be random variations in the rate of receipt of customers orders. Such variations can be accommodated by allowing a controlled level of smoothing buffer stock, and/or by a controlled variation of the output rate.

14.3 Process industries, seasonal products

14.3.1 *Wool*

On a large sheep farm, the sheep are shorn once a year. The wool obtained is cleaned, graded, weighed, and baled. It is then held in a store until orders are received and the bales are shipped to customers.

The size of the crop depends mainly on the number of sheep presented for shearing. The shearing is done by a visiting team of shearers, who may take two to three weeks to complete the work. For these operations, the main production control (PC) need is to maintain accurate records of the number of bales, in each quality grade, which are passed into stores, and of the number of sheep sheared by each shearer.

There is no further processing of the product on the farm after this stage. Accurate records are needed of the number of bales shipped and invoiced, and of the stocks remaining but there is little advantage to be gained by using Period Batch Control (PBC), or any other formal method of production control, to regulate the flow.

14.3.2 *Potatoes*

Potatoes are another highly seasonal product. The crop is lifted by potato harvester when it is ready, and is stored on a concrete floor in a large barn.

On one Scottish farm, seventy-six per cent of the crop goes to supermarkets, who require a washed, high-quality product bagged in polythene, in a range of different potato sizes (large, medium, small, and mixed), and quantities, or pack sizes (small, medium, and large), at a high call-off frequency.

Unlike the wool crop mentioned above, most of the processing of the potato crop (cleaning, inspection, grading, bagging, etc.) is best done when the supermarkets *call-off* deliveries from the stock, giving the mix of packs they require. In this case, PBC is useful to regulate the work.

The standard period is one day and the standard schedule is two days (see Fig. 14.2). The call-off instructions from the customers are received by fax. Orders received each day are packed and delivered on day 2. To make this possible the packaging units carry a small amount of smoothing stock of the most popular package sizes, making for stock when the orders are below the capacity level, and using this stock to supplement the output when the order level is above the capacity.

At 5.30 p.m. each day, the call-off list for next-day packaging is closed. All products called-off during the day up to this time are listed. Orders for other customers and for the farm shop are added. The total is assessed for load, and smoothing-stock orders will be added if necessary. The final list (programme) is given to the foreman of the packaging unit, to enable him to regulate the next-day's work.

Day number	1	2	3	4
Cycle 1	Ac	P		
Cycle 2		Ac	P	
Cycle 3			Ac	P

Key:

Ac = Accumulate orders

P = Pack and deliver

Packaging programme				Date:	
Order and call-off		Package			
Number	Customer	Pack	Size	Quantity of packs	
12070	A. B. Smith & Co	S	M	40	
12070	A. B. Smith & Co	M	M	45	
12070	A. B. Smith & Co	L	S	40	
13172	G. McBain	M	M	36	
13172	G. McBain	M	L	40	
13172	G. McBain	S	L	38	
13172	G. McBain	L	M	20	
13172	G. McBain	L	S	20	

FIG. 14.2 PBC for potato processing.

The advantages of PBC in this case are that it gives:

(1) a short lead time for the customer;
(2) an even load of work per day in the packing unit;
(3) reliable deliveries to customers.

14.4 A process industry with an even demand

Examples of this case can be found in many simple chemical factories, in cement mills and in mine-head installations for the treatment of ores.

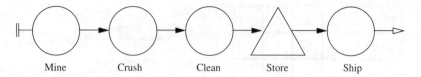

FIG. 14.3 PBC for ore treatment at a mine.

In most of these cases, the markets they feed require delivery on demand. To achieve this, the factories are obliged to hold some *delivery-buffer stock*. A standard-delivery-buffer-stock level is selected, deliveries to customers are made from stock on demand, and the factory makes at least enough of the product each period to replace these deliveries and to maintain the buffer-stock level.

Statistical analysis of the weekly, or daily, orders received is used to set the output level. The objective is to have enough product in stock at the beginning of each period to meet any likely period demand. There may be an occasional abnormal period demand which is much higher than average. If it is discovered that, say, x tons of stock will be enough to cover ninety five per cent of the period orders, and that twice as much stock ($2x$ tons) would be needed for one hundred per cent cover, then it may be better as a policy to adopt the lower level of cover, and to reduce the investment in stocks by a substantial amount.

14.4.1 Ore treatment at the head of a mine

Figure 14.3 shows the processes carried out in an ore-treatment installation at an iron-ore mine. This installation crushed the ore from the mine; removed earth and other residues by washing, settlement, sieving, and other processes; stored the cleaned ore; and finally loaded it into rail wagons to meet orders.

The mining and ore treatment are continuous, running at the rate needed to meet average demand. The main manifestation of imbalance between production and demand is found in the growth, or decline of the stock of the finished product. If the stock rises, this indicates the need for increased sales effort, and it may also indicate the need to reduce production. If the stock falls, this indicates a need to increase production.

Apart from the requirement for systematic statistical forecasting, it is doubtful if PBC or any other formal method of production control would be worthwhile in this case.

14.5 PBC in industries making products with a short shelf-life

As an example of a factory of the process-industry type making a food product with a short shelf-life, consider the manufacture of pork pies. Table 14.1 shows the range of products produced by a particular factory, and the percentage of total sales for each of them.

Table 14.1 The range of products and the percentage of annual sales

Identifying letter	Pre-description	Annual sales (%)
I	Individual pies	86
B	Small – buffet size.	6
R	Rectangular – large.	5
O	Oval – large .	3

The market is continuous throughout the year, with some seasonal variations.

14.5.1 *Pork-pie manufacture*

The main materials used are pork carcasses (deep frozen), flour, fats, and herbs. The carcasses and fat can be refrigerated and stored for long periods. There are economic advantages, however, if these stocks of materials are turned over rapidly to keep the stock-investment low.

Sales are made by delivery vans which visit food shops, hotels, and institutions (such as schools, hospitals, nursing homes and factory canteens) on a regular daily visiting round, with morning and afternoon rounds to different customers. They collect orders each day, normally for delivery on the following day. If stocks accumulate in the factory however the vans may carry additional stocks for direct sale, preferably to new customers.

The processes of production for the manufacture of pork pies are illustrated in Fig. 14.4. Three lines are in parallel, first meat cutting and the preparation

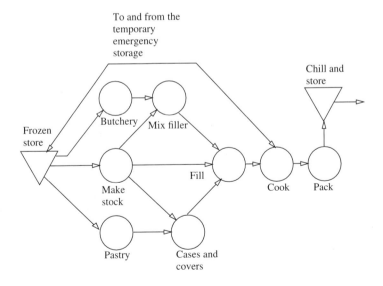

FIG. 14.4 The manufacture of pork pies.

(a) Standard ordering schedule (one day periods)

Day number	1	2	3	4	5
1. Cycle 1	Ac●	M	D		
2. Cycle 2		Ac●	M	D	
3. Cycle 3			Ac●	M	D

Key:

Ac = Accumulate
M = Make
D = Deliver
● = Programme meeting

(b) Van driver's order sheet

Round: South East		Week: 36	Day: Monday			Date: 6th Sept
Customer		Die type				Quantity
		I	B	O	R	
1	Post-office shop	12	35	–	–	47
2	Hospital	26	–	–	–	26

Key: I = Individual, B = Buffet, O = Large oval, R = Large rectangular.

FIG. 14.5 PBC for pork-pie manufacture.

of the pie-filling mix, second making stock, and third pastry preparation and the making of pie cases and covers. The output of these lines feed a final line which fills cooks cools chills and packages the pies.

The capacity of the factory is based on the average sales demand, with some special increases to cover the peak demand before national holidays. In normal periods, the factory will use not more than one day of the potential chilled shelf-life; but, before major holidays, this may grow to not more than two days. Minor changes in the capacity level are made by varying the overtime hours. Larger changes up to the limit of the machine capacity can be covered by contract labour or by additional employment.

The factory output is also planned using statistical methods, but the strict limitation of stocks due to shelf-life limitation means that it must be possible to change the production rate at short notice. The flexibility needed for this is obtained by:

(1) regulation of the number of pig carcasses released from frozen storage;

(2) the possibility of limited frozen storage of pies after cooking;

(3) direct sale of surpluses by delivery vans.

Carcasses are released from frozen storage in quantities which are each sufficient to provide one mixing batch of pie filler. The trouble is that the filler solidifies with time, and it cannot then be used for further production. Ideally, the number of batches per day should not be less than four for a full-days output. Once carcasses have been defrosted, they must be converted into *filler* and be used immediately to fill pies. The production output can be varied, however, by increasing or decreasing the number of batches of carcasses issued per day. Some further flexibility is also possible by freezing some pies for temporary storage.

14.5.2 *PBC for pork-pie factories*

The output level for the factory and the input of materials are regulated by statistical analysis. Due to the necessity for a strict regulation of stock levels, however, a quick-acting and over-riding control is needed to regulate delivery to customers and to react to random and special variations in demand. This is provided by PBC which essentially regulates the issue of frozen carcasses, the use of frozen-pie storage, and direct sales.

The period selected was one day, and the standard schedule was three days. See Fig. 14.5. The accumulation of orders each day is based on the van drivers order sheets. See again Figure 14.5. Their sales are totalled at the end of the day to find the total sales, and this value is compared with the selected production rate.

If the day's sales are less than the production rate, continuing at that rate will produce additional unsold pies. Possible action in such cases includes:

(1) distribution of the surplus pies to vans for direct sale;

(2) diversion of some pies to frozen storage;

(3) reduction of the number of carcasses issued, and of the output programme for the following day;

(4) And if the fall in sales persists, reduction of the production rate and reduction of the materials input.

If the day's sales are more than the present production rate, this may result in lost sales. Possible action in such cases include:

(1) drawing on frozen-pie stocks to complement the standard output;

(2) addition of one more issue of carcasses and work overtime;

(3) and, if the rise in sales persists, an increase in the capacity and the production rate, and an increase in the materials input.

At present, these decisions are made by a senior executive of the company. It will be noted that the van driver's order list for a round, subject to any additions for direct sale, provides the distribution list for the following day.

14.6 PBC in square industries

Square industries are similar to process industries in that the number of varieties of both materials and products are the same, but in the case of square industries these numbers are both large. Square industries are service industries. The materials they work on are supplied by the customer and are returned after processing.

14.6.1 *The dyeing of wool yarns*

As an example, a factory dyeing wool yarn for a number of spinning and weaving factories will be examined. The customers do their own carding and spinning, and they send the spun yarn to the dye mill for dyeing.

The skeins of yarn received are booked in, given a batch number, and tagged for identification. The dye shop is organized in groups, which are equipped to complete any orders they receive, and each batch is assigned to a particular group and to the next day's dyeing schedule in the sequence in which they are to be processed.

One problem is that the dye vats have to be cleaned at frequent intervals. The most difficult case is that of changing from a dark colour (for example, black) to a light colour (for example, light blue). Group organization helps by assigning very dark and very light colours to different groups, but the sequence of dyeing can still have a significant effect on the capacity of the dye shop.

After dyeing, the yarn is dried, inspected, washed, and is then returned to the customer.

14.6.2 *PBC for the control of dyeing*

In one dye mill, one-day periods are used for PBC, with a four-day standard schedule (see Fig. 14.6). Yarns received on day 1 are scoured and dyed on day 2, washed and dried on day 3, and returned to the customer on day 4.

Day number	1	2	3	4	5	6	7
Cycle 1	Ac	D	W	S			
Cycle 2		Ac	D	W	S		
Cycle 3			Ac	D	W	S	
Cycle 4				Ac	D	W	S

Key:
Ac = accumulate orders and list, D = dye
W = wash etc, S = distribution

FIG. 14.6 PBC for dyeing of yarn.

14.7 Summary

Process industries are simple industries in which a small number of material items is used to make a small number of different product types. Processing usually requires a small number of operations which are always done in the same sequence.

Under these conditions, statistical controls can be reliably used to predict the future output and capacity requirements, and to schedule the necessary input of materials. This, coupled with some control of product stock levels, is, in many cases the only system needed to regulate production.

Two cases discussed in this chapter indicate the types of product where PBC can be used with advantage in process industries; that is, products with:

(1) a high call-off frequency, short lead times, and many different types of package (for examples potatoes);
(2) a short shelf-life (for example, pork pies).

In both cases, the number of product varieties is near the limit normally indicated by the term *process industry*, and in both factories the number of product varieties is approaching the level where the factory might be classified as an implosive industry. In the case of the pork-pie factory, it was planned to increase the product range by adding Scotch eggs, paté, and sausages, making the factory even more implosive in type, and making PBC ever more desirable.

In the example from the square type of industry (a dyeing mill), the large variety of types of yarn and of different colours made the use of PBC desirable as a means for regulating production and controlling deliveries.

It should be noted that the standard period chosen for PBC in these industries is normally one day. This is possible with the process and implosive industries because very few different material items are needed, and it is possible with the square industries because the customer supplies the materials, and there is no problem of ordering materials. Shorter periods of half a day, or a shift, are seldom used except in food industries, where shelf-life is critical.

It is only in the explosive industries that the large number of bought material items makes it necessary to use the longer one week period, to allow an acceptable lead-time for call-off from suppliers.

15

PBC IN JOBBING FACTORIES

15.1 Introduction

Jobbing factories are those factories which manufacture special products. There is a low probability that any jobbing order will ever be repeated, and manufacture for stock is therefore impossible. Jobbing products may be components, assemblies, or complete assembled products. They may be ordered as single items or as a batch. In most cases, however, jobbing orders are for single items or for small quantities. Jobbing factories tend to specialize in a limited range of processes because it is impossible to make everything. For this reason it is just as easy to use Group Technology (GT) in a jobbing factory as it is in a factory making standard products.

Most jobbing orders require some preliminary *make-ready operations* to prepare them for production. Different orders may require different combinations of these make-ready operations, and they may need widely different lead times for completion. For this reason, in most jobbing factories the lead time for make-ready is estimated first, and this estimate is used in the calculation of the selling price and of the delivery promise date. When the make-ready operations have been completed, the orders are moved to a *ready-file*. Period Batch Control (PBC) is then used to regulate production.

15.2 Make-ready

The first task with a jobbing order is to plan the make-ready operations. In an engineering factory, for example, these may include different combinations of the following:

(1) the preparation of design drawings;
(2) the preparation of process routes;
(3) the manufacture or purchase of special tools (patterns, dies, form tools, etc.);
(4) the purchase of materials;
(5) software for NC machining (NC is numerical control of machine tools by computer);
(6) parts lists for assembled products.

The overriding objective must be to complete the make-ready in the shortest possible time. It will be noted that some of these operations are sequence dependent: the first operation has to be completed before the next operation can start. Traditional methods of doing this work assign each of these operations to a different department. This greatly inflates the throughput time for the total task, and in most companies it is more economical to create a single, centralized, make-ready department, which sees whole make-ready jobs through from start to finish.

The first task when an enquiry is received is to estimate, for tendering purposes, the throughput time and the cost of make-ready. Make-ready starts when an order is received. When *make-ready* has been completed, the order is moved to a *Ready file*, and PBC regulates production from that point on.

With complex assembled products with long and complex make-ready operations, it may be possible to reduce throughput times of the total order by starting some processing jobs before the completion of make-ready. This complicates PC, and it should only be used where it is possible to make significant savings in the throughput time. One problem is that it is relatively easy to bring forward the processing of simple parts, but these tend to have short throughput times, and bringing them forward does not reduce the total throughput time. It is more difficult to advance the work on the complex parts which have longer throughput times.

15.3 Planning PBC for jobbing

The first task in planning PBC is to choose the standard period, and to plan the standard programming and ordering schedules. These decisions must be made together. Because it is essential that each processing stage can be completed in one period, the division into stages selected for the standard ordering schedule must be known before the standard period can be chosen.

The objective should be to find the minimum-length period and the minimum number of processing stages with which reliable production is possible. This will reduce both the investment in stocks and the stock-holding costs.

The number of stages in the standard ordering schedule depends on the type of industry. The ideal would be one stage. Reasons for division into more than one stage include:

(1) incompatibility of processes, foundry stage for example, followed by a precision machining stage.

(2) component processing in several different groups, at a first stage all feeding the same assembly group, or groups at a following stage.

(3) one process supplies several groups at a following stage, for example, power saw for cutting blanks from bar.

(4) one process takes parts from several groups, for example painting or electroplating. (Fig. 15.1 illustrates a number of standard programming and ordering schedules used in different types of jobbing factory).

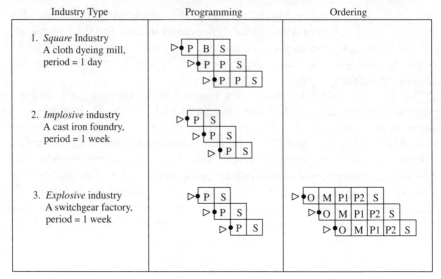

Key:-
● = programme meeting, P1 = production (plate weld), P2 = final assembly
▷ = from make-ready, O = obtain materials, M = make parts
S = distribution, P = production

FIG. 15.1 Standard programming and ordering schedules for jobbing.

The choice of the standard period must be such that it allows time to complete all the parts, including those with the longest throughput times, through any stage, assuming that there is sufficient capacity in machine-hours and man-hours to complete the work. If one week periods are chosen there is a maximum of 168 hours (7 × 24) available, and a minimum of, say, 55 hours per period (40 + 15 hours of overtime). If a company has GT, it must be making an exceptionally complex product if it cannot operate with one-week periods.

The key lies with reliability, particularly with respect to quality and plant maintenance. There are still some companies which add a ten per cent scrap allowance to their manufacturing orders. That is they are prepared to accept 100 000 quality rejects per million parts produced. There are factories in Japan operating with fewer than ten quality rejects per million. In these companies no scrap allowances or buffer stocks are needed; occasional overtime can be used to replace any scrap. Companies which have not yet achieved 1000 or fewer rejects per million—one reject per thousand—should not attempt PBC until they have improved their quality performance.

The same applies to plant maintenance. A well-organized maintenance system based on preventative-maintenance inspections, is essential for successful PBC.

15.4 Programming

In jobbing production, the annual programme is mainly a forecast of sales in money units or SUs—of weight volume, area, etc.—used as a basis for financial planning. It has less importance for PC in jobbing than is the case with other types of production.

In jobbing production, PC starts with the short-term production programme. Each programme meeting considers a list of outstanding orders which are *ready* for production—those for which make-ready has been completed—and orders are added to the production programme for the cycle, up to the limit of the available capacity (see Fig. 15.2).

In framing this programme, an attempt is made to complete all the orders by their promise dates. To help achieve this objective, the short-term sales programme will normally list all the orders not yet issued to production which are due for completion by the end of the cycle or which are overdue. It will be realized that the close connection between the short-term sales and production programmes in the case of standard products does not apply to jobbing products, due to the make-ready stage.

A method is needed for estimating the assembly load in man-hours, and the load in machine-hours on the critical machines. Without this, it is impossible to ensure that the load is less than the available capacity, and that the work to be done at each stage in each cycle can all be completed in one period.

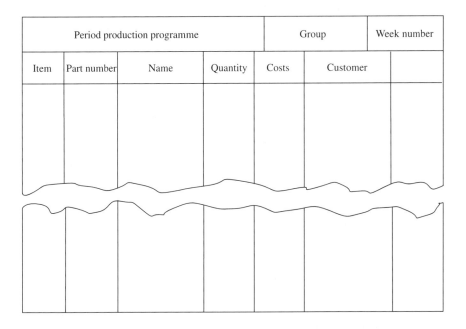

FIG. 15.2 A jobbing production programme for a cast iron foundry.

15.5 Ordering

If the factory is in a process, implosive, or square type of industry, the product is either bulk material or parts. In this case, the short-term programme lists the items to be made.

If the factory is in an explosive type of industry making an assembled product, make-ready will have produced a parts list (bill of materials) showing the parts and materials required to make the product, and indicating whether they are to be bought or made. If the company is organized with GT, make-ready will have already allocated each part to an existing group. A *list order* will now be issued to each group as soon as the short-term production programme is approved, showing the parts to be made in the cycle. It is uneconomical with explosive jobbing products to estimate the load on all the machine tools used. It is advisable, however, to estimate the load on at least the one or two most heavily loaded machines.

The ordering of purchased items will normally be treated as a part of the make-ready process, and the orders will not be programmed for production until all, or most, of the purchased parts have been received. If there are a few bought parts which are only needed at a late stage of assembly, and delivery has been promised by the supplier, it may be economical to start the production cycle before they are received.

15.6 Examples from practice

A few examples from practice will show how PBC can be used for jobbing in different types of industry.

15.6.1 *PBC in a small jobbing foundry*

A small jobbing foundry is an implosive type of industry. Make-ready is mainly concerned with the manufacture of patterns and core boxes. One-week periods, with a two week standard programming schedule (illustrated in Fig. 15.1), have been selected. The new programme for each cycle is prepared at the programme meeting on Friday each week, listing the orders ready for production up to the capacity limit for the foundry, measured in tons of castings. The load and the capacity are based on the weight of castings produced. The weight per casting is estimated during the make-ready stage.

This foundry also makes some castings in batches on call-off, for local factories. This work is done by a separate jolt-moulding group in the foundry. It is an advantage that the same PBC system can be used to regulate both types of production.

15.6.2 *PBC in a dyeing mill*

This is a square type of industry. The dyeing mill receives bolts of cloth from local weavers, dyes them to specified colours, and returns the dyed cloth to the weaver. See also the dyeing of yarn described in Chapter 14. The make-ready process is mainly concerned with the preparation of the dyes. It can normally be completed in one or two hours, but it may take longer if new dyes have to be purchased.

One-day periods have been selected, with a standard programming schedule of three days (illustrated in Fig. 15.1). The daily production programme takes the form of a simple list of orders which are ready for production. This list is limited by the known daily capacity of the mill. Both the load and the capacity are measured in units of yards of cloth.

A complication in this industry is the need to clean the vats after each order. Cleaning for a change from a dark colour to a light colour has to be more exacting, and it takes longer than cleaning for a change from a light colour to a darker colour. The company reserves different vats for light and for very dark colours, and it attempts to schedule the work by starting with the lighter colours in each range and moving progressively to darker colours, up to the stage when a major tank clean is needed.

15.6.3 *PBC in a factory making special switch gear*

This is an explosive jobbing industry. Special switch gear, for controlling the motor drives on special machines for the food-canning industry, and occasionally for other industries, are made in the factory.

The products This switch gear consists of switches, fuse units, instruments such as voltmeters, resistances, electronic controls, time switches, and other equipment, mounted inside a metal cabinet and wired together to regulate the starting, stopping, and operation of special machine tools. All these items are purchased. A range of four different standard cabinets is produced. It is only the equipment and circuitry which are special to each order.

Manufacture The flow of materials in switch-gear manufacture is illustrated in Fig. 15.3. The switch-gear cabinet and its doors are made in a GT group equipped with a guillotine, a universal punching and notching machine, a brake press, welding equipment, and a paint-spray cabinet.

Another GT group makes the mounting plate for the switches and the other equipment used. It contains a CNC (computer numerical control) punching machine, used to punch the mounting holes for all the different items to be mounted in the cabinet. The software programme for the CNC punching machine is prepared as part of the make-ready process. It is (CAD/CAM) application concerned with both design and manufacture. A third group machines various rotational parts.

A fifth GT group assembles the mounting plate assembly and other items into the cabinet; fixes certain switches, warning lights, and other items on the outside of the cabinet; it completes the wiring; it tests the whole unit, and it cleans and paints the unit ready for despatch.

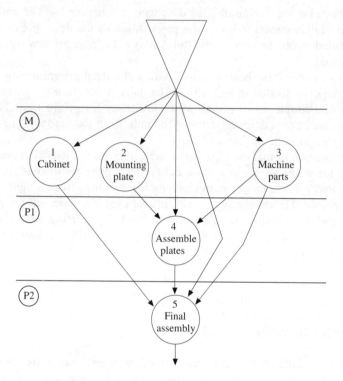

FIG. 15.3 The materials-flow system for a switch-gear factory.

The PBC system One-week periods have been selected, together with the five-week ordering schedule illustrated in Fig. 15.1. The short-term production programme is prepared on Friday afternoon each week by listing the orders that are ready, giving priority to overdue orders, and to orders due for delivery in four, or fewer, weeks time.

The number of sales orders in each programme is limited by the capacity available for CNC punching of the mounting plates, and for assembly and wiring of the mounting plates. It has been found that these are the critical processes, and that it is unnecessary to make detailed estimates of the load for the other processes.

The make-ready information given to the factory for each order consists of:

(1) a parts list;
(2) the layout of the mounting plate;
(3) software for the CNC punch, based on (2) above;
(4) the switch-gear wiring diagram.

This, together with the production programme, gives the factory manager all the data needed to regulate production.

15.7 PBC with large one-of-a-kind projects

The term *one-of-a-kind* or (OKP) is used for large jobbing projects, making products such as ships, bridges, dams, aeroplanes, and large buildings; but it is also used even when some of these types of product are made to standard designs, in quantity.

It is difficult to use PBC to control all the production in these cases, due to the long and different throughout times needed for different parts of the work. It is possible, however, to plan a master schedule to control the major stages in production, based on project-management methods, and to use PBC to regulate the manufacture of most of the details.

In ship building, for example, a master schedule can be prepared to regulate:

(1) the laying of the keel;

(2) manufacturing of the hull;

(3) installing machinery;

(4) launching;

(5) fitting-out;

(6) trials.

A difficulty for PBC is that, apart from taking varying periods of time, many of these stages overlap in time.

Shipyards will make many of the details they need, and for this purpose they may have:

(1) a machine shop;

(2) a sheet-metal shop;

(3) a cold-forging and a welding department;

(4) a pipe shop;

(5) a carpenter's shop.

At various stages in the building of a ship, orders will be sent to these departments for the manufacture of parts and assemblies. Inside these departments, manufacture can be regulated with advantage, using GT and PBC in the same way as they are used for smaller jobbing products in other explosive industries.

15.8 Summary

Period Batch Control can be used efficiently to regulate the production of jobbing products. Jobbing factories make special products. There is a low probability that any jobbing order will be repeated. They must be made to order, and there is no possibility, therefore, of making for stock.

Most jobbing orders need some preliminary *make-ready operations* to prepare them for production. It is necessary to estimate the cost and time required for make-ready, before a quotation can be given for jobbing work. As these times vary greatly between orders, it is difficult to include make-ready in the ordering schedule for PBC. The general solution is to regulate make-ready and production separately. When make-ready is finished, the job is moved to a *ready-file*, and production is controlled by PBC from that point.

Period Batch Control can be used for jobbing products in implosive, square, and explosive Industries. In the first two industries, the products are components, and the period production programmes list the items to be made each period. In explosive (assembly) industries, make-ready produces a parts list; and at a second stage after programming, orders are issued to groups for the parts to be made.

PART 4

INTRODUCING PBC

16

INTRODUCING PBC

Planning a new system is a relatively straightforward task. The implementation or introduction of the new system in a factory, is usually more difficult. This chapter looks at the problem of introducing Period Batch Control (PBC) in a factory, and suggests that the most efficient method for tackling this task is to divide the work into a series of short, independent, self-standing projects, each of which contributes to the total change and also gives some immediate benefits in its own right.

This chapter looks first at the problem of introducing PBC in a company in an explosive type of industry, which already has Group Technology (GT). It then looks at the same problem in other types of industry.

16.1 Introduction

The materials-flow system (MFS) of an enterprise is the system of routes between the places where work is done on materials to convert them into products. The change from traditional *process organization* (or functional organization) to *product organization—continuous line flow* (CLF) or *Group Technology* (GT)—is a change from an extremely complex MFS between organizational units which specialize in particular processes to a very simple MFS between organizational units (groups) each of which contain a variety of different work centres and is able to complete all the parts in its own particular *family* of parts or assemblies. Figure 2.1 illustrates this transformation.

One of the advantages of this change to GT is that it makes it easier to employ the simple single-cycle production control (PC) method of PBC, in place of the more complex and less reliable multicycle methods. Among other advantages, this change eliminates the *surge effect*. It also makes possible the more rhythmical and reliable regulation of the flow of materials through the MFS, which can be obtained with PBC.

The earlier chapters of this book studied the ways in which PBC is planned. This chapter looks at the methods used to introduce the change in a factory. It studies these methods progressively, under the following headings:

(1) objectives;
(2) preliminary changes;
(3) planning the series of change projects in explosive industries;

(4) planning the change projects in other industries;
(5) managing the change.

Methods for controlling the change schedule and expenditure are examined in the next chapter.

16.2 Objectives

The primary objective of PBC is to provide a simple and reliable method for regulating the flow of materials through the MFS of an enterprise. Period batch control is a single-cycle system, based on a standard period and on a standard programming schedule; or on a standard ordering schedule in the case of explosive industries. (See Fig. 16.1.)

The secondary objectives of PBC are:

(1) to eliminate the surge effect;
(2) to minimize stocks.

1. PBC is a Single-cycle system with common order and due dates (OD and DD respectively) in each cycle for all parts

Period number	1	2	3	4	5	6	7
Part number 1	OD	DD.OD	DD.OD	DD.OD	DD.OD	DD.OD	DD.OD	DD
Part number 2							
Part number 3							

2. It is based on a Standard period for example 1 shift, 1 day, 1 week, or 2 weeks.
3. It is also based on a standard programming schedule

Period number	1	2	3	4	5
Cycle number 1	Ac	P	S		
Cycle number 2		Ac	P	S	
Cycle number 3			Ac	P	S

Key: A/c = accumulate P = production S = sales

4. This is extended in explosive industries, to provide a standard ordering schedule also.

Period number	1	2	3	4	5	6	7
Cycle number 1	A/c	O	M	P	S		
Cycle number 2		A/c	O	M	P	S	
Cycle number 3			A/c	O	M	P	S

Key: O = get material M = make parts

FIG. 16.1 What is PBC?

INTRODUCING PBC

Although PBC has been used successfully with process organization, it is much easier to install, and it gives greater savings, if it is based on product organization—either CLF or GT.

It is desirable to restrict the change objectives to the introduction of the PBC system. Some companies have attempted to combine this change with other optimizations, such as the minimum cost of production and the maximum machine utilization. These additions tend to introduce conflicting objectives, and they so complicate the change that there is a risk that nothing worthwhile will be achieved.

This chapter starts by studying the introduction of PBC in an engineering factory, making a range of air compressors, which had already changed to GT. Other types of industry are then considered later in the chapter.

The company used as an example, employs a total of 325 personnel. It makes a range of small, medium, and large air compressors for builders, civil-engineering contractors, and industrial users. Some of the compressors are incorporated in portable compressor sets for use on building sites.

About half of the parts needed to make the compressors are *made* and half are *bought*. The made parts are produced in the GT groups illustrated in Fig. 16.2.

16.3 Preliminary changes

Period Batch Control introduces a number of special problems which do not have the same significance with multicycle PC systems. In particular, PBC requires major reductions in some component throughput times, in some operation set-up times, and in purchasing lead times. It is possible to take preliminary action to reduce stocks, in anticipation of the lower level of stock needed with PBC. This will release capital to finance the change.

16.3.1 *Reducing throughput times*

Group Technology itself reduces the throughput times for most made parts. Only a few exceptional parts will require special action to reduce their throughput times to the level needed to use PBC with one-week periods.

One way to find the critical items for which throughput time is likely to be a problem is to calculate the throughput-time index for all the parts made in a group. This is the available time (say a single-shift, that is, 40 hours plus 10 hours overtime) divided by the sum of the operation times for all the operations used to make a part. Ratios of two or less may require special action to reduce throughput times.

Possible methods to reduce throughput times for the critical parts, include:

(1) improving the group layout;

(2) adding shifts;

(3) reducing the number of operations per part (process integration);

(4) close-scheduling.

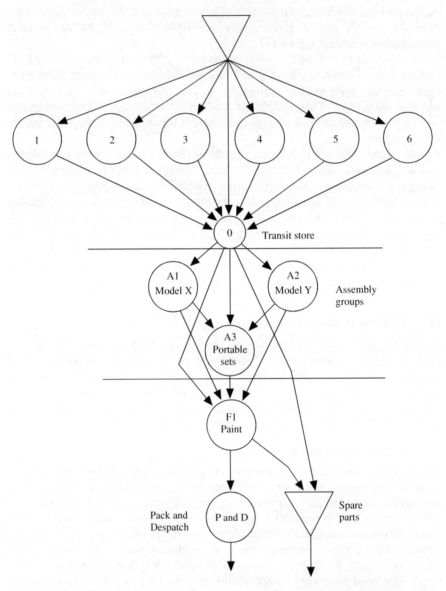

FIG. 16.2 GT groups in a compressor factory. The component groups are: (1) beds, (2) shaft, (3) gears, (4) pyrimidic, (5) small rotational, (6) sheet metal.

16.3.2 *Reducing set-up times*

A company which introduces PBC with one-week periods may (allowing two weeks for holidays) have to set up machines for some parts fifty times a year.

If such a part was previously made in batches six times a year, the machines must now be set up 8.3 times as frequently. If, however, the product in which the part is used only appears in the programme 17 times a year, the set-up frequency rises to only 2.8 times the present rate. If, however, the machine is only loaded for say thirty three per cent of the available time, it may be possible—depending on the ratio of the set-up time to the operation time—to absorb the additional setting-up time.

The simplest methods for reducing set-up times when this is necessary are *make-ready* and *sequencing* (see Chapter 9). Group Technology and PBC provide the conditions which make it possible to reduce set-up times for most operations to the level needed to operate PBC with one-week periods. Once again, it is only necessary to identify the few critical machines and parts for which additional set-up time reduction will be needed before PBC can be used with one-week periods.

A list should be prepared of machines with long set-up times and of the parts they make. Methods such as presetting, the integration of operations, dedicated machines for long operations, the standaridization of materials and design features, and process modifications can then be used to reduce the set-up times. The Toyota company in Japan has no machine in its large factory which requires even 6 minutes to reset. Set-up-time reduction has already been achieved: it can be done again.

16.3.3 *Reducing purchase delivery lead times*

The reduction of purchasing lead times was mentioned as one of the special needs of PBC. This requires the introduction of the call-off method, which needs the prior introduction of short-term programming. These are major PBC-introduction projects which will be discussed later.

16.3.4 *Reducing stocks*

One of the objectives of PBC is to reduce the stock investment, and to reduce the cost of stock holding. It is possible to take preliminary action to reduce stocks and to bring them into balance in product sets. This will release capital to finance the introduction of PBC, and it will also both simplify its introduction and reduce the risk of obsolescence. Two such methods are described below:

Eliminating nonrunners It is necessary to eliminate the risk that parts are being made which will never be needed. It is sometimes found that such parts are being made in a factory even though they will never be used because some processing machines in the factory are short of work and materials are available, or because operations on these parts pay high bonus rates to the workers.

One method of eliminating this risk of making parts which will never be used is to find any parts in stock which have not been sold or issued to assembly in

1.	Parts in the stores list	10 172
	Parts in current parts lists	4236
	Other parts sold as spares	260
	Other parts still needed — mainly new products in research and development	1377
2.	*Total current parts*	5873
3.	*Nonrunners*	4299
	Transfer nonrunners to suspense file	

FIG. 16.3 Correcting the databank. Note, records for current parts must be kept up to date, records for non runners are only corrected if they are resurrected.

the previous two years. Alternatively, starting with a list of all the different parts held in stock or listed in the stores list, any parts in current parts lists, or any other parts made in the last two years for sale as spare parts, can be removed to find the remaining *nonrunners*. All such items (see Fig. 16.3) should be transferred to a *suspense file*. All special patterns, dies, or other special tools used to make these parts should be moved to a locked store. Parts with no future should be scrapped, and no more of the parts should be made, nor should their special materials be purchased without high-level authority.

One advantage of this change is a considerable saving in the cost of servicing the data bank. Every time a new machine is installed, or an old machine is sold, and every time an operation is moved from one machine to another, the data bank must be corrected. If this is not done, the computer output will tend to be rubbish. If however, some parts are seldom, if ever, used, transferring them to a suspense file will eliminate the need to keep their processing-method records permanently up to date. The records will only need correction if the part is later brought back into production.

Getting stock into balance If a company has a multicycle PC system, and if it uses the pseudoscientific economic-batch quantity (EBQ) approach to fix batch quantities, the stocks in the stores of parts for an assembled product may vary between, say, nine months supply of some items and one-day's supply of others. There is a very high risk of obsolescence in such a situation. Action should be taken quickly to get the stocks into approximate balance. If this is not done then most of the parts required later, in the early weeks of introducing PBC, will come from stocks. It will be difficult to find work for the factory.

One way of balancing the stocks is to fix an arbitrary maximum batch frequency for all parts, of, say, six batches per year, each with a two-month order quantity. All parts with more than a two-month supply are taken out of production until their stocks drop to say a one-month supply. Remembering that, with the present case, allowing two weeks for holidays, one-week periods represent an order frequency of up to fifty runs per year for some items, the arbitrary change to six runs per year will only achieve the first rough balancing of the stocks. It may be advisable to make a further attempt to rough balance the stocks by reducing the order frequency, as a second step, to twelve batches per year, before starting with PBC.

16.3.5 *Eliminating intermediate subcontracted operations*

One change needed before introduction of PBC is the elimination as far as possible, of all intermediate subcontracted operations. Subcontracting the manufacture of complete parts is important as a method for balancing the load and the capacity. The subcontracting of single operations from the middle of a process route, however, is very difficult to control, and it tends to inflate throughput times. It should be avoided if possible.

Lists should be made of all intermediate subcontracted operations which could be made in the factory. If this work is returned to the factory, the additional load may help to maintain employment during the difficult time, when high existing stocks of made parts are being run down.

16.3.6 *Improving the plant layout*

Experience with GT has shown that it is not enough just to install the machines in a group together inside the same boundary. Attention must also be given to the layout of those machines normally used in the same sequence, to permit close-scheduling, with materials transfer in single units. Again, there is a need in many GT layouts, for provision for tool storage and presetting in the groups, in order to obtain the major set-up time reductions possible with make-ready and presetting.

Experience in Japan has shown the importance of notice-boards in the groups, to give group workers some information about their performance. Experience in Scandinavia has shown again the importance of facilities in the groups for rest and refreshment, to improve morale and job satisfaction.

These types of change will make it easier to introduce PBC, and they can be introduced as preliminary projects, before starting on the main changes.

16.4 Planning the series of change projects in explosive industries

Any major system change is best introduced as a series of independent change projects, so designed that each change moves the total project a step nearer to completion, and so that each change is independent and can be treated as a separate project for implementation. There is a need to find both the division into projects and the sequence in which they must be completed.

In the case of the compressor factory, the change was divided into the eleven projects listed in Table 16.1.

Table 16.1 Division into projects

(a) Preliminary projects

Project number	Project description
1	Reduce throughput times
2	Reduce set-up times
3	Reduce stocks
4	Eliminate intermediate subcontracting
5	Improve GT plant layout

(b) Introduction of PBC

Project number	Project description
6	Choose the period and the standard ordering schedule
7	Improve the annual programming
8	Introduce short-term programming
9	Introduce call-off purchasing
10	Introduce PBC ordering of made parts
11	Introduce Feedback control

16.4.1 *Preliminary projects (projects 1, 2, 3, 4 and 5)*

These projects were considered earlier in this chapter. Their objective is to release capital to help finance the changes, and to simplify their introduction by balancing the stocks into product sets. These preliminary projects are sequence independent, unlike the projects described in Sections 16.4.2–16.4.6, which are sequence dependent.

16.4.2 *Choosing the period and a standard ordering schedule (project 6)*

This must be the first of the introduction projects because the same division into periods is used for both the annual programme and the series of short-term programmes.

For process, implosive, and square industries, only a standard programming schedule is needed. With explosive industries, a standard ordering schedule is also needed. It is better in the case of explosive industries to start by planning and introducing the ordering schedule, even though it allows more time than is necessary during the early changes in programming.

16.4.3 *Improving annual programming (project 7)*

Changes will be needed in this programme, if only to make its output reconcilable with the later series of short-term programmes.

16.4.4 *Introducing short-term programming (project 8)*

New short-term programmes for sales, production, and stocks are required at the beginning of each period. In the case of the compressor factory, the ordering schedule includes periods for *obtaining materials, making parts, assembly*; and *delivery* to *customer*; and programmes are needed for sales and for production.

When first introduced, the short-term programmes are used solely to regulate assembly. The regular routine of programme meetings and programme planning are established over several weeks, before going on with projects 9–11.

16.4.5 *Introducing the call-off method (project 9)*

The introduction of short-term programming provides the foundation needed for the call-off method, because it provides an accurate indication of when parts will be needed for assembly. It is desirable to start with Pareto analysis to find the purchased items with the highest annual value, and to bring together other parts requiring the same processes for purchase from the same source.

In addition to selecting the parts with which to start the call-off method, it is also necessary to analyse the performance of suppliers and to decide where to start with the introduction of the method. Some suppliers will see the call-off method as an imposition. It helps if a company can offer additional business (by reducing the number of sources) and can offer shorter credit terms for payment as incentives.

Generally, a company starts with some buffer stock of bought parts. If the supplier performance is carefully monitored and controlled, however, it should eventually be possible to eliminate the need for a store to hold finished bought parts. They can be delivered to the point of use when needed.

16.4.6 *Introducing an ordering and loading system for made parts (project 10)*

Each group in the factory needs, at the beginning of each period cycle, a list of all the parts or assemblies it must produce by the end of the period, and a *load summary* showing the load in machine-hours, and man-hours imposed by this list order on all the machines and other work centres in the group. Lists of bought parts per product and their suppliers are also needed, so that the quantities per period can be calculated from the programme, to be called-off from each supplier. Because these quantities of parts and loads are directly proportional to the number of products in the short-term programme, the list orders and load summaries are very easy to produce.

The first task in this project is to produce the ordering standards for each product and to design the simple computer software needed to use this data with the short-term production programmes, to produce list-orders and load summaries for the groups, and call-off instructions for suppliers (see Fig. 13.3).

The only problems likely to arise with this project are those due to parts with critical throughput times or to operations with critical set-up times. Such problems can be solved if the effort is made.

16.4.7 *Introducing feedback controls (project 11)*

The system outputs from each group need monitoring coupled with feedback control, so that the performance of the groups can be kept under continual review. The outputs which it is desirable to monitor include:

(1) the output of made parts;
(2) the inspection rejects per million parts;
(3) the overdue orders;
(4) the capacity and the load;
(5) the absenteeism;
(6) the plant idle time.

16.5 Planning the change projects in other industries

16.5.1 *Process industries*

Process industries convert a few varieties of material into an equally small number of varieties of product, and because they normally use a small number of processes, always in the same sequence, they are normally organized for 'Continuous Line Flow' (CLF).

All the projects listed in Table 16.1, except projects 4, 5, 9, and 10, may be needed, but the work to be done is much simpler than is the case with the explosive industries. The ordering of materials is particularly simple because it is restricted in process industries to a very small number of varieties; this helps to make statistical forecasting reliable.

In most types of production, periods of one week are best, particularly for the initial introduction. The process industries, however, include many food-processing companies, where low *shelf-lives* make it necessary to use much shorter periods of a day or a shift. This may complicate the introduction of PBC.

16.5.2 *Planning the change for implosive industries*

Implosive industries convert a small number of material varieties into a large number of products. The ordering of materials is again very simple, as in the case of the process industries.

The products in this type of industry are parts. Their production is regulated by the issue of short-term programmes, which list the items to be made, and consist mainly of firm sales orders. All the projects listed in Table 16.1 except projects 4, 9, and 10, may be needed. The introduction of PBC is more complicated than for process industries, but it is less complex than for explosive industries.

16.5.3 *Planning the change for square industries*

Square industries convert a large number of material items into an equally large number of product items. The materials are normally provided by the customer, and the products are these same items after processing.

This is the simplest case for the introduction of PBC. All the projects listed in Table 16.1—except projects 3, 4, 9, and 10—may be needed, but generally in a simpler form than is required in the explosive industries.

The main advantages of PBC in these industries is that the regular rhythm of periodic control improves the delivery performance, improves accountability, and increases job satisfaction.

16.6 Summary

The introduction of PBC is best achieved by a succession of independent change projects, each of which advances the completion of the total change, and gives some benefits in its own right. Firstly, there are a number of preliminary projects, which will make it easier to introduce PBC. These projects are usually independent. They can be introduced in any required sequence, and will give benefits even if PBC is not introduced. Next, there are the PBC-introduction projects which are sequence dependent, and must be introduced progressively in one special sequence.

The explosive (assembly) industries provide the most difficult case for PBC. The process, implosive, and square industries make bulk, or piece-part products, and the introduction of PBC is simpler. The division into projects is similar, but in these cases there is no progressive component-processing stage, and the product output is regulated by the short-term production programmes.

There is a need to plan and control the introduction of the change, in respect of both the schedule for introduction and the cost and benefits of the change. This aspect of the work is studied in the next chapter.

17

CONTROLLING PBC INTRODUCTION AND OPERATION

17.1 Introduction

Control in management is the process which constrains events to follow plans. It works by monitoring the actual values for significant output variables at regular period intervals, by comparing these values with planned values for the same variables, and by feeding back information about important variances between them to the appropriate manager so that corrective action can be taken.

It is an important, if pessimistic, principle of management that nothing works unless it is controlled. This chapter looks at the controls needed for both the introduction of PBC and for its operation during production.

17.2 Controlling the introduction of PBC

The main plans which have to be controlled during the introduction of PBC are:

(1) the schedule for introduction;
(2) the investment for introduction;
(3) the savings induced by the change.

17.2.1 *The schedule*

It is necessary to plan the progressive division of the work of introducing PBC into independent projects, and to fix target dates for starting and finishing each of them. Next, a routine method is needed for monitoring the actual performance, and to ensure that failure to meet the target dates is not ignored. This should lead to some action.

It is important however, to realize that this change to PBC is part of a major cultural change in industry. Failure to meet a target may indicate a need to amend the schedule to give time for further discussion and training rather than a need to enforce immediate compliance.

17.2.2 *The investment for introduction*

It is also necessary to budget and control any special expenditures required to implement the change. These will normally be concerned with investment or

CONTROLLING PBC INTRODUCTION AND OPERATION 233

nonrecurring expenditure on such items as a new computer computer software, new furniture, new printed forms, the employment of a consultant, and other similar outlays.

17.2.3 *Savings and benefits*

Finally, it is also necessary to control the net savings, or the sum of all changes (plus or minus) in the investment, and in the operating expenditure. To do this it is first necessary to estimate the future savings. Total accuracy is not possible. The estimate must, however, be between the highest figure which might be achievable in practice and the lowest figure which would make the change financially attractive.

The principle change in investment, with the introduction of PBC, is likely to be in the stock investment. It is the potential reduction in the stock investment which is most frequently given as a reason for introducing PBC.

The main savings in operating expenses come from reduced throughput times, leading to lower stocks and stock-holding costs, lower materials-handling costs, lower set-up times and costs and lower numbers of quality rejects. There is no advantage to be gained by considering expenditures which will not change. If, for example, there will be ten people in production control (PC), both before and after the change, there is no change in the salary expenditure. If there are differences they are best treated as savings (plus), or losses (minus) in the operating costs. Any changes in the operating costs are best treated in this way.

There are other benefits from the change to PBC (such as better accountability improved morale, and increased job satisfaction), which are difficult to quantify, and also, therefore, difficult to control in terms of money. Periodic management audits can be used to assess the value of these types of benefit.

17.3 The schedule for introduction

The project must be started by dividing it into a set of stand-alone projects, each of which will provide some benefits on its own, and will also advance the total project a stage towards completion. An example from an engineering company making a range of assembled mechanical products found the projects listed in Table 17.1.

Table 17.1 Projects for the introduction of PBC

(a)   *Five preliminary projects*

Project	Project description
p1	check the accuracy of the data bank
p2	reduce and balance the stocks
p3	reduce the set-up times
p4	improve the group layout
p5	eliminate the subcontracting of intermediate operations

Table 17.1 (*contd*)

(b)	Twelve main projects
Project	Project description
m1	plan the standard period and ordering schedule
m2	revise the annual programming system
m3	plan the period programming system
m4	introduce period programming to regulate assembly
m5	plan the call-off method of purchasing
m6	introduce the call-off method for *bought* parts
m7	introduce the call-off method for *bought* materials
m8	plan the PBC ordering system for *made parts*
m9	introduce the PBC ordering system for *made parts*
m10	introduce periodic *loading*
m11	introduce *feedback controls*
m12	plan the first-operation loading sequence for LSS operation scheduling in each period, in each group

17.3.1 *Preliminary projects*

The preliminary projects are designed to facilitate the main project of introducing PBC. First, it is desirable to check the accuracy of the data bank, and to make any necessary corrections (see Chapter 5). Initial changes to reduce and balance stocks should release capital to help finance the change, and they should also simplify the introduction of PBC. If the introduction of Group Technology (GT) has not already eliminated all intermediate subcontracted operations, this should be done at this point, before introducing PBC.

With regard to setting-up-time reduction, it is probable that simple low-cost methods such as *make-ready* and *sequencing* will achieve eighty-five per cent or more of the reduction needed to introduce PBC with one-week periods. Apart from planning and introducing these two methods of set-up reduction, all other parts and machines where the set-up time may be a problem should be identified, the planning of the reduction of these times should be started.

Finally, remembering that all parts must be completed at each stage, in a single period, the group layouts should be improved as far as possible, to reduce throughput times. This may involve such changes as: first, moving machines which are normally used in the same sequence so they are close to each other to form cells inside the groups, in order to make *close-scheduling* possible; and secondly storing tools in the groups, to obtain more easily the savings in set-up time which are possible with the *make-ready* and *presetting* of tooling for set-up changes.

17.3.2 *Main projects*

One-week periods were selected with the four-week ordering schedule shown in Fig. 17.1. Shorter periods than one week might have been possible, but this would have increased the total set-up time and reduced the lead time offered to suppliers. The main projects are self-explanatory from their descriptions in Table 17.1. They are listed in the sequence in which they need to be carried out.

CONTROLLING PBC INTRODUCTION AND OPERATION

Week number	1	2	3	4	5	6	7
Cycle 1	Ac	◆ M	A	S			
Cycle 2		Ac	◆ M	A	S		
Cycle 3			Ac	◆ M	A	S	

Key:
Ac = accumulate orders, ◆ = programme meeting,
M = make parts, A = assemble, and S = distribution.

FIG. 17.1 A selected ordering schedule.

The most significant change comes with main project 4 the introduction of period programming. This not only provides a flexible method for regulating assembly, but it also provides both the information needed for call-off purchasing and the foundation for ordering made parts with PBC.

Main projects m10, m11, and m12 are concerned with the feedback controls to be used during operation, and with the information needed for operation scheduling. They deal first with loading or checking the load against the capacity, next with the feedback controls of *progressing* (to ensure that the actual output matches the planned output by the required due dates, with *inventory control* (to ensure that the stocks stay within the required limits), and finally with the information needed for operation scheduling.

Operation scheduling with GT and PBC is very simple; it is normally delegated to the group leaders. It may be desirable, however, to preplan the best sequence for loading first-operation machines at the beginning of each cycle, so that in the loading of later operations, it can be left to queuing discipline to keep the bottleneck machines loaded and to keep tooling families of parts together. This method of operation scheduling is known as *launch-sequence scheduling* (LSS).

17.3.3 *Critical-path analysis of project scheduling*

Critical-path analysis (CPA) can be used to find the critical path, between those projects which determine the total time needed for the change to Period Batch Control (PBC). Figure 17.2 shows the critical-path analysis chart for the case described earlier in this chapter.

The Gantt-chart schedule illustrated in Figure 17.3 is based on the information gained from the CPA. This chart can be used not only as a plan, but also as a record of achievement in meeting the plan.

17.4 The investment in the change to PBC

The investment needed will depend on what facilities are already available. If the company now uses Materials-Requirement Planning (MRP), for example,

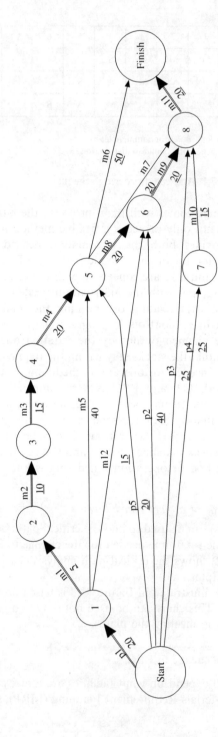

FIG. 17.2 Critical path analysis (CPA) for introducing PBC.

Key: (1) See Table 17.1 for the project key (for p and m).
(2) The duration times are underlined and they are measured in days (5 days = 1 week).
(3) ⟶ Critical path, 130 days or 26 weeks.

Project		Week number																									
Number	Name	1	2	3	4	5	6	7	8	9	10	11	12	13	14	15	16	17	18	19	20	21	22	23	24	25	26
p1	Data bank																										
p2	Stocks																										
p3	Set-up time																										
p4	Group layout																										
p5	Subcontract																										
m1	Period end schedule																										
m2	Annual programme																										
m3	Period programme																										
m4	Introduce period programme																										
m5	Plan ca l-of																										
m6	Call-off bought parts																										
m7	Call-off bought materials																										
m8	Plan ordering made																										
m9	Introduce ordering made																										
m10	Introduce loading																										
m11	Introduce controls																										
m12	LSS list																										

FIG. 17.3 A Gantt-chart schedule for introducing PBC.

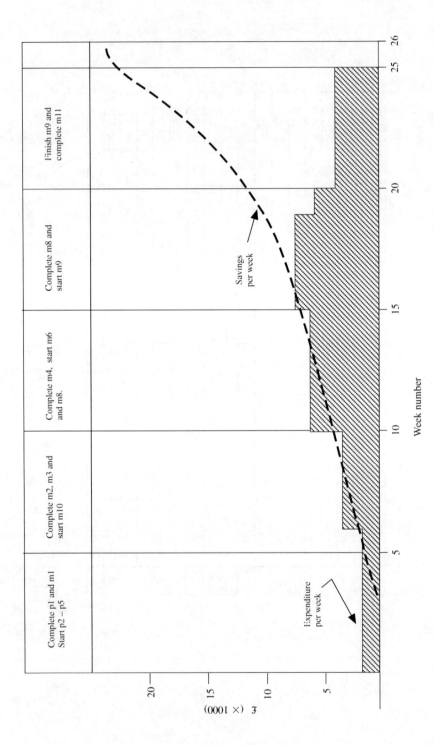

FIG. 17.4 An estimate of the expenditure and the savings.

the change will release more computer capacity than is required for PBC, and it will be possible still to use a small part of the existing software.

It is necessary to plan exactly what facilities will be needed for the change. Next, it must be determined which of these facilities are already available, and then the cost of the remaining facilities must be estimated and their delivery must be scheduled. One of the objectives when introducing PBC is to make the change using existing resources, without requiring any additional capital. It will often be possible to finance the change from the reduction in the stock investment. Figure 17.4 shows special expenditure and savings forecast over the twenty-six-week scheduled introduction period.

17.5 Savings and benefits

The savings and benefits which can be obtained by introducing GT are listed in Fig. 2.5. The additional savings obtained by introducing PBC are listed in Fig. 17.5. In order to control whether these savings are actually obtained, it is necessary to forecast, for each period, the savings which should be achieved—in order to provide a target—then to measure at the end of each period the savings actually obtained and finally to determine the difference between the plans and the achievement—to show where action is needed by management.

Figure 17.6 illustrates the type of project control chart needed. Only the economic advantages of the change are considered because they can be evaluated in numerical terms and are easily measured. The advantages of any social benefits are difficult to express in numerical terms, and are much more difficult to measure. An attempt should be made, however, at regular intervals during the project, to determine if social advantages are being gained.

In addition to the savings obtained with GT (see Fig. 2.5), PBC gives the following additional savings:

1. A reduction in the run quantities, due to a common run frequency, giving:
   (a) a further reduction in the throughput time,
   (b) lower stocks,
   (c) lower stock-holding costs,
   (d) better customer service.

2. More efficient load/capacity control, due to the period load all being completed in one period.

3. Savings in kitting-out for assembly (parts are made in this period for assembly in the next period).

4. Savings in storage cost (parts are only made when they are needed — JIT).

5. A reduction in materials obsolescence (parts are always made in balanced product sets).

6. An elimination of the *surge effect* due to single-cycle ordering.

FIG. 17.5 The advantages of PBC.

		Start	Target	Week number										
				2	4	6	8	10	12	14	16	18	22	26
Throughput time (weeks)	Max	14	2	12	12	10	9	9	8	7	6	4	2	2
	Min	2	2	2	2	2	2	2	2	2	2	2	2	2
	Mean	10	2	9	9	7	7	7	6	5	4	3	2	2
Set-up time (min)	Max	300	20	150	140	140	135	130	100	80	50	30	25	20
	Min	12	–	12	8	8	5	2	2	–	–	–	–	–
	Mean	155	8	81	72	72	70	65	50	40	25	15	12.5	8
1. Bought call-off (%)		2	100	2	2	3	3	3	4	4	11	38	66	92
2. Sales (× 1000 £)		4200	4200	4200	4200	4200	4200	4200	4220	4220	4230	4230	4230	4230
3. Stock (× 1000 £)		1400	350	1400	1350	1203	1000	933	808	531	455	423	425	360
4. Stock turns per year		3	12	3	3.1	3.49	4.2	4.5	5.2	7.9	9.3	10	10	11.75
5. Overdue sales		28	–	12	–	20	18	–	10	5	–	5	–	–
6. Overdue made parts		63	–	49	30	33	30	27	23	12	2	–	1	2
7. Overdue purchases		29	–	30	36	27	22	14	4	5	3	1	–	2

FIG. 17.6 A control chart for introducing PBC.

If it is found that the advantages gained are less than was expected, an attempt should be made to determine why this is so, and to recover the missing benefits.

17.6 Control during operation

When the change project has been completed and PBC is installed and running, the control used to ensure that the project achieves its target has fulfilled its purpose, and it can be discontinued. A monitoring system is now needed to ensure that any gains which are achieved are maintained, and to ensure that the system continues to operate efficiently.

Control data is needed to indicate to the management how the groups and the production control system are performing. Information is also needed by the people working in the groups, so that they can gauge how well they are doing in following the production plans.

Figure 17.7 shows an example of a control chart. A chart is prepared in an engineering company for each group in a machine shop. The chart is exhibited on a notice-board at the entrance to each group, and it is brought up to date each Monday morning.

17.7 Summary

There is a need to control the introduction of PBC in a factory to ensure that the required objectives are achieved. The main plans which have to be

Group control chart (1994)								Group name:				Group number:	
Week number	1	2	3	4	5	6	7	8	9	10	11	12	13
1. Overdue parts													
2. Excess parts													
3. Rejects per million													
4. Workers in group													
5. Absentees (days)													
6. Overtime (hours per week)													
7. Excess hours													
8. Machine down time													
9. Number of set-ups per week													
10. Late materials													

FIG. 17.7 A group control chart.

controlled are: the schedule for introduction the investment for introduction and the savings induced by the change.

When the introduction has been completed, there is still a need for regular control of the performance in operation, first to guide management, and secondly as feedback information for those who work in the groups.

18

CONCLUSION

18.1 Introduction

Since the start of the Industrial Revolution, production control (PC) has been the weak link in production, and the task of regulating the flow of materials through factories has not yet been mastered.

The introduction of computers into industry after the Second World War brought some benefits to PC. The main advantage of computers has been the elimination of clerical drudgery. Their main disadvantage is they have made it possible to use very complex and, unfortunately, also very inefficient systems, which suffer major distortions in performance because they fail to compensate for dynamic influences such as the *surge effect* and Jay. W. Forrester's *industrial-dynamics effect*.

Until recently, industrialists with PC problems would often be told that their problems were due to their computer software. They were told that if they invested £300 000 (sterling) in new Materials-Requirement Planning (MRP) software, and if they then changed their operating system to suit the software, all their problems would be solved. The results obtained by following this kind of advice have been disappointing.

A new philosophy is gradually gaining acceptance. This sees the key to efficient production in a simple materials-flow system (MFS), involving the change from process organization to continuous line flow (CLF) or Group Technology (GT). Simplification of the MFS makes it possible to adopt the simple single-cycle PC method known as Period Batch Control (PBC). This change, among other advantages, eliminates the dynamic distortions caused by the surge effect, and it leads to much more reliable and predictable PC.

18.2 The materials-flow system

The MFS of a factory is the system of routes followed by materials between the places where work is done on them to convert them into products.

The MFS is always very complex in companies with process (functional) organization (as shown in Fig. 18.1). The change to GT greatly simplifies the system. If production-flow analysis (PFA) is used to plan the groups, it is always possible to make the change from process organization to GT so that groups complete all the parts they make without back-flow or cross-flow of materials between groups, and with few if any exceptional parts. It is generally

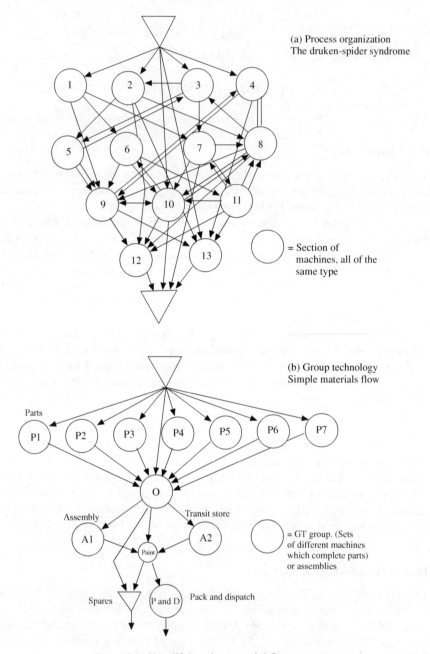

FIG. 18.1 Simplifying the material-flow system.

possible to make the change without altering the design of the products or the basic processing methods, and without having to buy any new machines. Such purchases may be desirable, but they are seldom, if ever, necessary as a prerequisite for the introduction of GT.

If PFA is used to plan the division into groups, this division is usually very stable. Future development of the processing methods and of the plant, and changes in the product design and range, can generally be accommodated without altering the basic division into groups.

The principle objective when planning GT is to find a total division into groups which complete all the parts they make, without back-flow or cross-flow between them. If this is achieved, the advantages gained include:

(1) a major reduction in the throughput time;
(2) leading to reduced stocks;
(3) and to a reduction in stock-holding costs;
(4) a reduction in materials-handling cost;
(5) improved accountability, because groups complete parts, the group leaders can assume the responsibility for quality, cost, and completion by due-date;
(6) increased job satisfaction;
(7) a reduced set-up time;
(8) an increased rate of return on investment.

These advantages arise because groups are small and all in one place under one leader; and because they complete all the parts they make. Hybrid groups, and other variants of GT which do not have these qualities, do not gain these advantages.

18.3 Simplifying production control

The simplification of the MFS makes it possible to simplify the PC system. The main changes needed to do so are:

(1) better quality and processing reliability;
(2) shorter throughput times (just-in-time);
(3) single-cycle ordering;
(4) reduced set-up times;
(5) reduced run quantities;
(6) reduced stocks.

18.3.1 *Better quality*

A company which plans a scrap allowance of ten per cent on its orders is allowing up to 100 000 rejects per million parts produced. This represents an enormous possible loss of materials, high storage costs, and a very low level of reliability. Efficient PC cannot be achieved with such poor reliability.

Group Technology itself contributes to improved quality, because parts are completed at each stage in a single group by one team of workers. Among other benefits, this makes it possible to improve accountability, which helps to improve the quality. Some engineering companies in Japan are currently operating with fewer than ten rejects per million parts. This is a reasonable and achievable target for quality improvement. The replacement of a few rejects at this level of reliability can be covered by occasional overtime. At this level, there is no need for a scrap allowance or for the holding of insurance-buffer stocks.

18.3.2 Single-cycle ordering

Most companies in the world still currently use multicycle ordering systems, ordering different parts with different order frequencies which are often based on the pseudoscientific economic-batch quantity (EBQ) theorem.

We must change to single-cycle ordering, if for no other reason than that this is the only known way to avoid the *surge effect* (see Fig. 1.8). Other advantages are that it eliminates the need for stores between processing stages, and that it also eliminates the need for *kitting-out* for assembly. If parts are received from suppliers, or if they are made each period in the exact quantities needed for assembly in the next period, then there is no need for kitting-out.

Single-cycle ordering needs to operate with short periods, in order to reduce stocks and to improve the rate of return on investment. This in turn requires short throughput times, short set-up times, and run quantities based on PBC standard periods.

18.3.3 Reduced throughput times

Group Technology itself normally reduces throughput times sufficiently to make it possible to use single-cycle ordering with one-week periods. If some critical parts are found to need a further reduction in throughput times, this can be achieved by process planning to reduce the number of operations per part, coupled with layout improvement in the groups. The close-scheduling of parts made in batches, and shift working, are other ways of achieving the same objective.

18.3.4 Reduced set-up times

Operating with short periods such as a week tends to induce an increase in run frequency up to fifty runs per year for some parts. This induces an increase in the total set-up time, which may reduce the capacity on some machines. Some reduction in set-up times will usually be needed for bottleneck and other heavily loaded machines, and for machines with long natural set-up times, to compensate for this increase.

Most of the set-up-time reduction needed to work with one-week periods in explosive industries, can be achieved with simple low-cost methods, such as make-ready, sequencing, the dedication of machines and the standardization of materials, tools, and design features. Further savings can be gained at some cost, by presetting, and by the redesign of tooling or of machine tools.

It should be remembered that in the very large Toyota car-manufacturing plant in Japan there is no machine which takes even ten minutes to set-up. Set-up-time reduction has already been achieved, so at least we know it is possible.

18.3.5 *Reduced run quantities*

A reduction in run quantities with an increase in run frequencies is one of the methods which can be used to reduce stock investment. The use of short periods in PBC has this effect. As already explained, such a reduction tends to increase the number of set-ups and to reduce capacity. It is necessary, therefore, to compensate for this loss by reducing the set-up times per set-up on some machines.

18.3.6 *Reducing stocks*

A primary aim of any modern PC system is the increasing or the rate of return on investment. By reducing both the investment (stocks) and by increasing profits (reducing stock-holding costs), a reduction in stocks induces a geared increase in the rate of return on investment. To reduce stocks, it is necessary to reduce throughput times, purchase-delivery batch quantities, and run quantities, as explained earlier.

18.4 Period Batch Control

Period Batch Control (PBC) is the method of production control which exploits the changes needed to simplify production control, using the methods described in the previous section.

This method starts by choosing a standard period for ordering, and then chooses a standard programming schedule and also, in the case of explosive industries, a standard ordering schedule to be repeated every cycle. Figure 18.2 shows as an example an ordering schedule based on one-week periods and four stages.

The shorter the period, the lower the level of stock needed to operate the system. The shorter is the standard ordering schedule, or series of ordering stages per cycle, the lower (once again), is the level of stock needed to operate the system.

Factors which influence the choice of the period include the throughput time for manufacture, the set-up time on the critical machines, the lead time acceptable to suppliers for call-off, and the shelf-life of the product. Periods of one week have been found to be acceptable in a wide range of industries, but shorter periods of, say, a day must be used in industries where the products have like food products a short shelf-life.

Factors which influence the choice of the standard schedule includes the incompatibility of successive processes, the reliability of suppliers, the shelf-life of the product, and the nature of the market.

1. Period Batch Control is a single-cycle ordering system, based on a standard period.

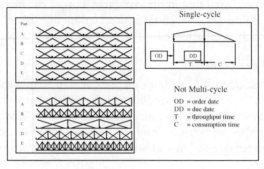

2. One week periods are common, but shorter periods of one day may be needed for food products. Longer periods have also been used.

3. Processing stages may be unavoidable for example, stages for machining and assembly. In this case standard programming and/or ordering schedules will be need for example a four period ordering schedule:

Week No	1	2	3	4	5	6	7
Cycle 1	O	● M	A	S			
Cycle 2		O	● M	A	S		
Cycle 3			O	● M	A	S	

Key: ● = programme meeting, O = obtain materials,

FIG. 18.2 Simplifying the regulation of materials flow with PBC.

The objective is always to reduce the stock investment, but the PBC system can be modified without difficulty to use smoothing stocks where necessary in order to save capacity, and to deliver from stock where this is essential in order to make sales.

18.5 Economic savings with GT plus PBC

It is not immediately obvious how the change to PBC simplifies production control, and leads to a more efficient production system, assuming that the MFS is based on GT. The main reasons can be listed as follows:

1. PBC eliminates the need for interstage stores between, for example, *component processing* and *assembly*. Parts are made and/or received from suppliers each period in the exact quantities needed for assembly in the next period.

2. PBC greatly simplifies *kitting-out* for assembly. If the exact number of parts needed for assembly in the next period can be made in the factory, or

can be delivered by suppliers in the current period, then the need for kitting-out is eliminated.

3. PBC increases the savings in set-up time obtained by sequencing. With PBC, all or most of the parts in any *tooling family* using the same tools and/or set-up on a machine will be on order together in the same period. With traditional multicycle ordering, only a few of these parts will be on order together.

4. Period Batch Control greatly simplifies operation scheduling. Instead of one large schedule for a department, smaller, and therefore simpler, schedules are needed, one for each group. Parts are ordered in sets at the beginning of each period for completion by the end of the period. If there is sufficient capacity, it is comparatively simple to find a schedule to complete the work.

5. PBC simplifies ordering. Most of the data needed to order production is directly proportional to the production programme. For example:

   (a) The number of products to be assembled in each assembly group, per period, is given in the short-term production programme.
   (b) The number of parts to be made in each component-processing group, per period, is found by simple explosion from the short-term production programme.
   (c) The amount of materials required per group, per period, is again found by simple explosion from the short-term production programme.

6. Period batch control simplifies loading, or the regulation of the load, to ensure that it is inside the limit imposed by capacity. The net load on any machine in a group is directly proportional to the numbers of each product in the short-term production programme which uses the machine.

Figure 18.3 provides a summary of the potential economic advantages of GT plus PBC. None of these advantages is automatic. It is necessary to take management action to achieve them. For example, the advantages of improved accountability are not obtained until the responsibility has been officially delegated to the group leaders. Total GT plus PBC, however, makes it possible to achieve savings such as these, which cannot be obtained with traditional process (functional) organization and multicycle PC.

18.6 The social benefits of GT and PBC

In addition to the economic benefits of GT plus PBC there are also social benefits, such as improved morale and increased job satisfaction. Such benefits are generally ignored by accountants because they are difficult to quantify. There are, however, several managers who have successfully introduced GT and/or PBC, who see the main advantages gained, in the change in workers' attitudes.

(a) The advantages of GT over process organization:

(1) *short throughput times*, because machines are close together, giving lower stocks and stock-holding costs and better customer service,
(2) *better quality* (fewer rejects), because groups complete parts and the machines are close together under one foreman,
(3) *lower materials-handling costs* because the machines are close together,
(4) *better accountability*, because groups complete parts and the foremen are responsible for quality and cost and for completion by due date,
(5) *training for promotion*, the group is a mini factory,
(6) *automation* is simpler to introduce because a group is an FMS with some manual operations,
(7) *reduced set-up time*, because similar parts occur together on the same machines,
(8) *greater job satisfaction*, because more people prefer to work in groups and to finish parts or products.

(b) Additional advantages of GT plus PBC:

(1) *reduced run quantities*, giving lower throughput times, stocks, and stock-holding costs, and giving better customer service,
(2) *better load/capacity control*, because period load is all completed in one period,
(3) *elimination of kitting-out*, because parts are made and received from suppliers in the current period for assembly in the next period,
(4) *elimination of the stage stores*, because parts are only made when they are needed (JIT),
(5) *reductions in materials obsolescence*, because parts are always made in balanced product sets,
(6) *elimination of the surge effect*, because PBC is a single-cycle system.

FIG. 18.3 The advantages of GT and PBC.

Some scientists with a physics or engineering back ground find it difficult to accept the findings of the behavioural sciences. A main problem is that it is difficult to use controlled experiments. In physics, for example, most experiments end by returning to the original conditions, to ensure that they have not changed. This type of controlled experiment is not possible with human experiments, because the study of human beings tends to change their attitudes and behaviour, making a *controlled* experiment impossible.

Although the physical and the behavioural sciences are different, there is considerable experimental evidence which indicates the conditions needed for improved job satisfaction and morale. Of these the following are among the most important:

1. People need to belong to groups with common aims.
2. They need to work together in teams.
3. They need to be associated with the completion of products.
4. They need some freedom to plan the way in which they work.
5. They need the possibility to vary their pace of work.
6. They need information about their achievements in meeting targets (feedback).
7. They need direct access to authority.

18.6.1 *Groups and teams*

Some of the early research in this field, was in the introduction of the *long-wall method* for cutting coal in coal mines. It was found that teams which worked together to clear a given coal face, were much more efficient than with an alternative method where different specialists visited the coalface in turn. Also, the workers preferred team working.

The popularity of team training in industry seems to indicate that this finding is widely accepted today. Group Technology is the ultimate expression of this belief.

18.6.2 *Completion of products*

Much research in the behavioural sciences has pointed to the importance of the role of completion of products in promoting job satisfaction. Most workers prefer to work in a group which completes parts to working in a section which specializes in only one process.

The change from process organization to Group Technology tends to increase job satisfaction.

18.6.3 *Freedom to plan the way of working*

Freedom to plan the way of working is not always possible. In high-technology industries, the method is enshrined in the process route, and in the choice of tooling and of computer software. It is mainly with manual processes that this freedom is relevant.

18.6.4 *Variation of the pace of work*

Physiological research has shown that human beings achieve their highest performance if they work hard for short periods, with rest periods in between. They do not achieve maximum efficiency if they attempt to work like machines at a fixed even rate. This need has been covered in several GT applications, by allowing a controlled accumulation of buffer stocks between assembly stages, to provide time for relaxation.

18.6.5 *Feedback*

Workers need information about their performance in meeting production targets. With GT and PBC this is normally provided by the regular monitoring of performance and the display of this information in groups.

18.6.6 *Direct access to authority*

The need for direct access to authority is partly the expression of a deep distrust of the alternative, bureaucratic, form of authority: better the devil you know than some faceless authority. With GT, authority is generally delegated to group leaders, and the workers in the groups do have direct access to authority.

18.7 Conclusion

Period Batch Control is a simple flow-control, single-cycle method of production control, based on a single, standard, planning period, and on a standard ordering schedule—linking a standard series of processing stages—which is repeated every cycle.

Period Batch Control is a approach which can be adopted to suit most types of industry. It has been used in process, implosive, square, and explosive industries. It has been used with continuous line flow, with batch production and with jobbing production, and it can be adaped for *smoothing* and for ex-delivery stock.

Period Batch Control is a general system, of production control like *stock control*—based on stock reorder levels—and Materials-Requirement Planning (MRP), it is applicable to most types of industry. It is not a special solution for a small number of special situations. It is essentially much more efficient than stock control, or MRP. One advantage of PBC in comparison with stock control and MRP is that it eliminates the surge effect. It is also a just-in-time (JIT) production control system which can operate efficiently at very high rates of stock turnover.

GLOSSARY OF THE TERMS USED IN GT AND PBC

Arterial-flow system A network diagram showing the flow of materials between suppliers, works, and customers, and the flow of money and information between these units, and between the functions of management.

Assembly group An organizational unit, which completes the assembly of a product or a major stage in its assembly, and which is provided with all the people and facilities needed to do so.

Automation The total mechanization of the processes of manufacture, including the automatic control of these processes. For continuous line flow, the automated form is the automatic transfer line. For Group Technology, the automated form is the Flexible Manufacturing System (FMS).

Back-flow Materials flow which goes from an early stage in manufacture to a later stage, and then flows back again, for example, from the machine shop to assembly and then back again to the machine shop.

Batch A generic term covering all the ways in which parts or assemblies are batched together for convenience in production. It includes the order quantity, the run quantity, the set-up quantity, the transfer quantity, and the delivery batch quantity. Batch is synonymous with lot in USA.

Branched groups A method of group assembly in which some groups complete different major assemblies, which flow on completion to another group which joins them together to complete the product.

Buffer stock An arbitrary addition to the stock of a component or product, held as an *insurance buffer* against scrap or late delivery, or as a *smoothing buffer* to smooth random or seasonal variations in demand, or as a *delivery buffer* to make it possible to give immediate delivery from stock to customers.

Call-off A method of purchasing in which the purchase contract does not quote delivery dates. The customer tells the supplier the quantities of each item to be delivered each period, by issuing *call-off instructions* against the contract at regular periodic intervals.

Capacity The number of *units of work* (in machine-hours or man-hours) which can be completed at a work centre in a given time period.

Cell A number of machines and other facilities in a group which are normally used together and which are laid out near each other to form a single work centre, which can be operated by a single worker or by a small team of workers. (Note that the term *cell* is sometimes used as a synonym for the term *group*. It is, however, an unfortunate word in this sense, because of the close association of the word *cell* with prison cells.)

Classification and coding (C&C) A method of classifying and coding parts based on their shape, function, or other design characteristics. It was often used in the past to find families of parts for Group Technology.

GLOSSARY

Company-flow analysis (CFA) The first subtechnique of production-flow analysis, which attempts to simplify the materials flow between the factories in large companies.

Continuous line flow (CLF) A simple form of *product organization* in which most products use the same processes in the same sequence and there is a continuous flow of materials between the processes. It is widely used in the simpler process industries, for assembly and for some components in mass production.

Close-scheduling This is the starting of a following operation on a part before the completion of the previous operation on all of the parts in a batch. It is only easy to use if the machines are close together, as with Group Technology.

Common machine A category of machine tools in the SICGE code, for which there are several of the type, and for which it is comparatively simple to transfer an operation done on one of these machines to another machine of the same type, or to some other closely related machine, for example, centre lathes, turret lathes, chucking capstans, and turning centres in a machine shop, may be closely related common machines.

Critical-path analysis (CPA) A technique based on networks, used in project planning, to find the critical series of processes, which limit the time needed to complete a project.

Cross-flow The flow of materials between different groups at the same processing stage. Not permitted with Group Technology.

Cycle stock The stock generated by manufacturing and purchasing in batches. It does not include buffer stock or excess stock.

Delivery time The elapsed time between the receipt of a purchase order and delivery of the goods to the customer. Note delivery lead time is a forecast of future delivery times.

Demand variation The way in which customer demand for products varies from period to period. It includes random, seasonal, and cyclical variations.

Dispatching Dispatching is the third level of progressive production control, following programming and ordering. It plans, directs, and controls the completion of shop orders in factory workshops.

Due date The date, or series of dates, by which an order is to be completed.

Economic batch quantity (EBQ) A discredited theory which postulates that for every item made, or bought by a company, there is one batch quantity, the EBQ, which will give the minimum cost. It imposes the use of different batch quantities for different components, requiring the use of multicycle ordering. (The term *economic-lot size* is used in the USA.)

Excess stock Stock in excess of planned stock (that is, in excess of the cycle stock plus the buffer stock).

Explosive industry A type of industry in which large numbers of components are combined to produce a relatively small number of products. It includes most assembly industries.

Families of parts Sets of parts which are similar in shape or function, or are all made in the same group. Also, tooling families, or sets of parts which all use the same set of tools on a particular machine.

Feedback controls Controls which operate by planning future values for system output variables at regular intervals, measuring their actual values, comparing the actual with the planned values, and feeding back information about significant variances between them to management so that action can be taken to steer future values back into line with the plans. (See *monitoring*.)

GLOSSARY

Factory-flow analysis (FFA) The second subtechnique of production-flow analysis. It seeks to find a division of factories into major-groups or departments which complete all the parts they make, and to find a simple unidirectional materials-flow system between them.

Flexible programming The progressive scheduling of a series of short-term (for example, one week) sales and production programmes, which are used as the basis for Period Batch Control. (The term *short-term programming* is a synonym.)

Flexibility of labour Labour is flexible if it has been trained to do a variety of different tasks.

Flow-control systems These are systems of production control which base ordering on explosion from production programmes. (For the opposing term see *stock-base ordering systems*.)

Flexible manufacturing system (FMS) An automated production system which can complete a variety of different parts in a variety of different run quantities. A group-technology group can usually be described as one or more partly automated FMSs.

Group analysis (GA) The third subtechnique of production flow analysis (PFA). After factory-flow analysis, it seeks to find a total division of departments into groups of machines and people, each of which completes all the items it makes, without back-flow or cross-flow to other groups.

General machines A category of machine tools in the SICGE code for which there are generally few of the type which are widely used in processing many different types of part, for example, saws for cutting blanks from bars and automatic painting plants for painting parts.

Groups The parts of a department forming the smallest organizational units with Group Technology. Groups contain a variety of machine tools (and/or other processing facilities) and workers with a variety of different skills. Groups should complete all the items they make, without material back-flow or cross-flow to other groups.

Group layout The layout of machines and other processing facilities on the floor of a workshop, to form a group.

Group Technology (GT) A type of product organization for manufacturing, in which a factory is progressively divided into departments and/or groups which complete all the items they make without back-flow or cross-flow between them. (Note, the term *Group Technology* was used by Professor Mitrofanov of Leningrad University as the title for his research into the relationship between the shape of parts and their processing methods. He showed that lathes could be set up to process sets of similar parts, thus saving set-up time. By adding drills and milling machines, groups were formed which completed parts. The meaning of the term evolved in this way to its present meaning as a type of organization.)

Groups in parallel A layout method for group assembly in which a number of different groups all make the same product.

Groups in series A layout method for group assembly in which a series of groups laid out as a line work progressively to complete products.

Implosive industry A type of industry in which a small variety of materials is used to produce a large variety of products, for example, foundries, glass works, potteries, etc.

Industrial dynamics The name of Jay. W. Forrester's research into the amplification effect, where the amplitude of demand variation tends to grow when demand is transmitted through a series of independent agencies, for example, from retailer to wholesaler to factory stock.

Inspection A feed-back control system used to maintain the quality of products, in which monitoring involves measurement of the work done and comparison of these measurements with the dimensions and tolerances specified in product drawings and specifications.

Intermediate machine A category of machines in the SICGE code for which there is more than one of the type which would have been classified as a special machine if there had been only one such machine.

Inventory control A feed-back control used to constrain the investment in stocks to stay within prescribed limits.

Jobbing production A type of production in which the products are special, being made as single units or in small quantities, against a single order. Jobbing orders may be repeated, but there can be no certainty that this will happen. Jobbing orders may or may not include the design of the product.

Just-in-time A method of production control in which products are only assembled when they can be delivered to customers; parts are only made when they are needed for assembly, or for delivery to customers; and the receipt of purchases is only accepted when they are immediately required for assembly or for further processing.

Kitting-out The collecting of sets of parts, ready for assembly.

Line analysis (LA) The fourth subtechnique of production flow analysis. It analyses the way in which materials flow between the machines in a group, to find information needed to plan the layout of the group.

Lead time A forecast of future *throughput times* for manufacturing, or of *delivery times* for purchases.

Load The amount of work—in machine-hours or man-hours—imposed by a given programme, or a given set of orders, on a work centre.

Loading The comparison of work load and capacity both in units of machine-hours or man-hours.

Launch sequence scheduling (LSS) A method of operation scheduling used with Group Technology and Period Batch Control, based on planning the sequence of jobs for the first-operation machines, in each group each period, coupled with processing in strict order of the job arrival sequence, for the other machines in the group.

Make-ready (1) A method used to reduce the set-up time. All materials, tools, drawings, etc. for the next job, must be ready for setting up by a machine before the previous job is completed. (2) Methods used to prepare jobbing orders for processing.

Manufacturing The making of wares by hand or machine. (Also see *production*.)

Materials flow system The system of routes followed by materials in a factory between the places where work is done on them.

Modules Modules are minimatrices formed progressively during group analysis. Each module includes all the parts not already allocated to previous modules which use a particular key machine. The key machines are used in the sequence dictated by the *special plant list*.

Monitoring The comparing at regular intervals of actual with planned values for a system output variable and the feeding back of information about major variances between them, to management for correction.

GLOSSARY

Monogroup A layout method for group assembly in which a single group completes the assembly of a product or set of products.

Materials-Requirement Planning (MRP) A flow control, multicycle, production control ordering method, in which requirement quantities are found by explosion from a production forecast for some months into the future. The scheduling of orders for parts is generally based on a fixed order quantity and/or lead time for each part.

Multicycle ordering Production control ordering in which different parts have different order frequencies. It is often associated with (so-called) *economic batch quantities*.

One-of-a-kind product (OKP) A one-of-a-kind large product such as a ship, a bridge, or a large building. The term is also sometimes used for a small batch of such products.

Operation scheduling A part of *dispatching* involving the scheduling of work operations on machines or other work centres, in workshops.

Order frequency A parameter giving the number of orders issued for a part per year.

Order quantity A parameter giving the number of parts or products authorized for manufacture by a shop order, or for delivery by a purchase order.

Ordering Ordering is the second level of progressive production control. Following programming, it plans, directs, and controls the workshop-ordering and purchase-ordering activities in a company.

Organization The division of an enterprise into organizational units (for example, divisions, departments, sections, or groups) and the assignment of responsibility for their management to people.

Parameter A system variable to which a manager can assign arbitrary values at will. For example: order quantity, run quantity, selling price.

Parts list A list of the parts needed to assemble a product. The term *bill of materials* is used in the USA.

Period batch control (PBC) A single-cycle, production-control method in which ordering is based on periodic explosion from a series of equal, short-term production programmes. It was one of the first just-in-time systems.

PBC period The standard period, chosen for a PBC system. *Purchase delivery quantity (PDQ)* The number of items in a purchase delivery.

Production-flow analysis (PFA) A technique for planning the change from process organization to Group Technology (product organization) and for the simplification of materials-flow systems. It comprises the following five subtechniques used progressively: (1) company-flow analysis, (2) factory-flow analysis, (3) group analysis, (4) line analysis, and (5) fooling analysis

Perfect flow project The name of research by the author, based on computer simulation at the ILO's Turin International Centre, which defined the *surge effect* and explained why it occurred.

Planned stock This is cycle stock plus buffer stock.

Plant layout Plans showing the position of machines in a workshop.

Presetting The setting up of tools in special holders, before setting-up starts on a machine, so that they can later be added quickly to the machine, where they will present the tools accurately in the correct position for machining. Presetting reduces the set-up time on the machine.

Primary network A *materials*-flow system network based on a small number of common process-route (numbers PRN), which are used, each to make a large number of different parts. It provides the starting point for the simplification of materials flow in factory-flow analysis. (FFA.)

Process industry A type of industry in which a small variety of materials is used to make an equally small variety of products, for example cement, milk, cheese, or bread.

Process organization A type of organization in which organizational units specialize in particular processes. (*Functional organization* is a synonym.)

Process route number (PRN) A code number showing the series of machine tools and other work centres visited by a part during manufacture.

Processing stage A stage in processing at which batches must be completed before forwarding them to the next operation, for example foundry, machining, and assembly stages.

Production the manufacture and distribution of goods.

Product organization A type of organization in which organizational units complete all the parts or assemblies they make, for example, continuous line flow and group technology.

Production control (PC) (1) The function of management which plans, directs, and controls the materials supply and processing activities of an enterprise. (2) The function of management which plans, directs and controls the flow of materials through the materials-flow system of an enterprise.

Production planning The function of management which plans, directs, and controls the physical means to be used to manufacture products in an enterprise.

Programming Programming is the first level of progressive production control. It schedules the delivery of finished products to customers (the sales programme), and the assembly and/or packaging of finished products (production programme). (The term *master production scheduling* is used in the USA.)

Progressing A feedback control used to constrain output to follow the plans made in programming; ordering of made and purchased parts; or in operation scheduling.

Random variations Minor variations on both sides of a mean (sales) rate which are statistically random by nature.

Reallocation of operations The reallocation of operations from one machine to another to eliminate exceptions when forming GT groups. This is a part of group analysis.

Run quantity A parameter giving the number of identical parts which are produced at a work centre before changing to make some other part.

Schedule (1) (verb) The planning of the starting and/or finishing times for work tasks. (2) A list or chart showing the planned starting and/or finishing times for work tasks.

Set-up quantity A parameter giving the number of parts, not necessarily all of the same design, which are to be produced at a work centre before the set-up is changed.

Sequencing The planning of the sequence of loading work on a machine tool so that parts in the same tooling families are machined one after the other in order to reduce the set-up times.

Shop order An order issued to a workshop requiring the manufacture of a specified quantity of parts or products by a particular due date or by a series of due dates. (*Workshop order* is a synonym.)

Short-term programming See *flexible programming*.

SICGE code A method used to code machine tools according to the quantity of each type available and to the possibility of transferring the work they do onto other machines. There are five categories: special(S), intermediate(I), common(C), general(G), and equipment(E).

Single-cycle ordering Production-control ordering in which orders are issued on a series of common order dates at equal time intervals throughout the year. All orders issued each period share a common order date and a common due-date.

Smoothing A reduction in the variation in production rate induced by random and seasonal variations in sales demand made by making additional products for stock when demand is below the mean sales demand rate, and using these stocks to supplement production when the demand is above the mean. This allows production to continue at an even rate, and reduces capacity losses.

Special machines A category of machine in the SICGE code where there is one only of the type, and for which it would be very difficult to transfer the work that it does to any other machine in the factory.

Special plant list (SPL) The plant list arranged in SICGE code sequence, with the machines in each code block arranged in the sequence of increasing numbers of the parts (F) they machine. It is used in group analysis to find the key machines for the formation of modules.

Square industries A type of industry in which a large variety of materials is used to make approximately the same number of finished products. The customer usually provides the materials, which are returned after processing.

Stock Stock is all the tangible assets of a company other than its fixed assets.

Stock-base ordering systems Production control ordering systems in which the release of orders is based on changes in the stock level. (Also see *flow-control systems*.)

Stock control An ordering system of the stock-base type in which the ordering attempts to maintain the level of stocks in a store. A fixed order quantity is ordered when the stock drops to a preset reorder level.

Subcontract The buying of an item which could be made in the factory. It is often done to reduce load and to increase free capacity.

Surge effect Large, unpredictable variations in stocks and/or load found with multicycle ordering when the peaks and troughs of different component stock or load cycles drift into and out of phase with time.

Tooling analysis (TA) The fifth subtechnique of production-flow analysis. Following after line analysis, it seeks to find *tooling families* of parts, which are all made using tools from the same set, and/or can all be made at the same set-up.

Tooling families See *families of parts*.

Transfer quantity A parameter giving the number of parts which are to be transferred as a batch between machines doing consecutive operations.

Work centre A set of machines and/or other processing facilities which are normally used together and are treated as a single unit for scheduling. (See *cell*.)

INDEX

accounting stocks 76–7
advantages, introduction of PBC 198, 233, 247, 248–9
agricultural machine factory, production control (example) 189–97
assembled products
 ordering standards 129
 programme schedules 111–18
assembly groups 34–6
 materials transfer 36–40
assembly load 53

batch buying 125–6
batch processes 170
batch quantity 16, 60–1
bills of material, see parts lists
bottleneck machines, loading 168, 173
buffer stock 58, 65, 67, 69–71
 smoothing 67–9
business forecasting 97

call-off method, purchasing 53, 126–7, 151, 229
capacity-load control 136–8
cast-iron foundries, see foundry production
change projects planning
 explosive industries 227–30
 other industries 230–1
changes, introduction of PBC 223–7
circulating capital 58
classification and coding (C&C) systems 29
cleaning services 162
close-scheduling 61, 141
connectance (variables) 44, 46
continuous line flow (CLF) 22, 125
coolant recovery processes 160
critical-path analysis (CPA) 235
cure-network diagram 46
customers lists 90
cycle stock 58, 60–1
 reduction 61, 65
cyclical variations, sales 104

database
 maintaining 91–3
 requirements 81–90
decorated-glass production control (example) 186–8
dedicated machines 149
delivery buffer stocks 69–71, 75
demand-magnification effect (stock levels) 74

demand variation effects 10
dispatching systems 3, 18–21, 152
due-date-filing system 19–20
dyeing mill, production control (example) 214–15

economic-batch quantity (EBQ) 11, 14–18, 126, 226
economic-lot size, see economic-batch quantity (EBQ)
employees list 84–6, 92
excess stock 58, 71
expediter, see progress chaser
explosive industries 125, 189
 classification 6
 materials flow system 190–1
 ordering standards 129
 planning change projects 227–30

factory-flow analysis (FFA) 31
feedback controls 3, 229–30
finished-product stock programme 106
flexible manufacturing system (FMS) 25
flexible programming 5, 14, 108–9
 jobbing products 121–2
 period selection 109
 piece parts 119
 process industries 122–3
 programme meeting 123–4
 schedule selection 109–11
 standard assembled products 111–18
flow-control systems 6–9
forecasting 97, 98
Forrester effect, see demand-magnification effect
foundry production 119–21, 214

Gantt chart scheduling method 20–1
GLOSSARY 253–9
group analysis (GA) 31
Group Technology (GT) 14, 18, 21, 24–5, 41, 61
 group leader authority 152
 introduction 234
 materials transfer 36–40
 savings and benefits 29, 221, 239, 248–9, 249–51
 scheduling 163–4
 stages 25–8
 stock levels 76
 throughput times 109, 141

INDEX

housekeeping, *see* cleaning services

idle time scheduling 165
implosive industries 125, 179
  classification 6
  decorated-glass production example 186–8
  decorative laminate production example 179–86
  flexible programming 119–21
  materials ordering 127–8
  planning change projects 230
inspection process 159
insurance stocks 65
interference effect (stock levels) 73
inventory control 3, 74
  for assembled products 75
  for piece parts 75–7
investment requirements 232–3

jobbing products 121–2, 210–14
just-in-case ordering systems 12
just-in-time (JIT) ordering systems 12, 14, 71, 108

Kanban system 9, 12
kitting out 18, 248

laminate production control (example) 179–86
launch-sequence scheduling (LSS) 21, 163
  component planning 173–4
  flexibility 174–5
  planning 164–70
  reliability 170–3
limestone quarrying 128
line analysis (LA) 31
loading control 3, 53–5
  simplifying 249
long-wall mining method 251
lot, *see* batch
lot size, *see* batch quantity

machine-tool factory, production control (example) 197–9
machining load 54–5
made parts, ordering and loading 229
make-ready tasks 121, 143, 153, 210
manufacturing industries, classification 6
market research 97
master scheduling, *see* programming
materials-flow system (MFS) 22, 221, 243–5
  explosive industries 190–1
  materials handling 160
  transfer 36–40

Materials-Requirement Planning (MRP) 8–9, 21, 135, 243
matrix resolution 31
modules, machine tool 32–4
moving annual total (MAT) method 98

next-job decision rules 21
nonrunners 225–6
nonstandard parts, *see* jobbing products

objectives, introduction of PBC 222–3, 247
one-of-a-kind (OKP) projects 217
operation route number 141
operation scheduling 18, 152, 163–4, 249
operation sheets 89–90, 93
optimum production technology (OPT) 12
ordering schedules 129–33
ordering systems 3
  multicycle 11–12
  simplifying 249
  single-cycle 11
order quantity 60
ore-treatment installation, production control (example) 204
overdue orders list 51–2

parts lists 88–9, 93
Period Batch Control (PBC) applications
  dyeing mill 214–15
  explosive industries 189, 197
  implosive industries 179, 186
  jobbing factories 210
  jobbing foundry 214
  one-of-a-kind projects 217
  pork-pie factories 207
  process industries 200
  short shelf-life products 204
  square industries 208
  switchgear factory 215
piece-part products 119–21
pig iron deliveries 128
planning, corporate 5
planning horizon, *see* programming, term
plant layout 227
plant list 81–4, 91–2
plant maintenance 160
plant-type code 82
pork pie factory, production control (example) 204–7
potatoes, production control (example) 202
presetting, tooling 156
problems, introduction of PBC 198–9
process (horizontal/functional) organization 22
process industries 200
  classification 6
  even demand 203–4
  materials ordering 127–8

INDEX

process industries (*cont.*)
  planning change projects 230
  programming 122–3
  seasonal products 202
  short shelf-life products 204–7
process integration 141
process routes 86–8, 92–3
production control (PC)
  database 81, 91–3
  definition 3, 22, 42–3
  developments 243
  simplifying 245–7
  production-flow analysis (PFA) 29–31, 154, 243, 245
production programmes, *see* programming, production
product life cycle 97, 104
product (vertical) organization 22
programme meeting 123–4
programme schedules 109–11
  annual 47, 49–51, 97, 106
programming 3–5, 94
  flexible, *see* flexible programming
  production 5, 6, 47, 49–51, 94, 106
  sales 5–6, 94, 97
  short-term, *see* flexible programming
  stock 5, 94, 106
  term (duration) 94
progress chaser 52
progressing control 3, 47–53
project control chart 239
purchase lead times 151, 225
purchase-order progressing 53

quality-control department 159

records, ordering process 162
requirement scheduling 6, 9
run frequency 60, 61
run quantity 60, 61
  reduction 141, 247

sales programmes, *see* programming, sales
savings and benefits, introduction of PBC 198, 233, 248–9
scheduling 3, 232
seasonal products 202
seasonal variations, sales 67–9, 102
sequence-constraint problem 149–50
sequencing process 144–5, 153–4
setting-up, machine 153–4
set-up quantity 60
set-up time 143–9, 168, 172–3, 224–5, 246–7
shift working 141
shop-order progressing 51–2
shortage-chasing system 52
short-term-production-programme controls 47, 49

SICGE code (machine selection) 33–4, 84
single-cycle system 222, 246
single minute exchange of dies (SMED) book 147
single-shot machining 149
single sourcing 126–7
smoothing stock 38, 75, 116
social benefits 249–51
special plant list (SPL) 34
square industries 125, 200
  classification 6
  planning change projects 230–1
standardization benefits, materials and design 145
statistical forecasting 97, 98
stock-base ordering systems 9–10
stock-control (SC) systems 9, 21, 74–5, 77, 135
stock programmes, *see* programming, stock
stock-ratio optimization (STROP) system 10
stock-replacement (SR) systems 9
stocks
  balanced levels 226–7
  definition 57
  delivery 116–18, 121
  dynamic changes 73–4
  excess levels 58, 71
  holding levels 225
  reduction 247
  types 58
subcontracted operations 172, 227
suppliers lists 90
surge effect (stock levels) 12, 73–4, 246
swarf removal 160
switchgear factory, production control (example) 215–16
system design 46–7
systems theory 43–4

throughput time 60, 61, 139–43, 223, 246
time constraints 135–6
time-cover (TC) systems 10
tooling
  development 148–9
  regrinding 159
  storage and maintenance 156–9
tooling analysis (TA) 31, 90, 154, 168
tooling banks 159
tooling families 168, 191, 249
transit store 38
trends analysis 98

wool dyeing factory, production control (example) 208
wool, production control (example) 202
wool spinning, purchases 128
work-in-progress stock 57